WILD EDIBLE UNDERUTILIZED PLANTS

Nutritional, Antinutritional, and Nutraceutical Aspects

WILD EDIBLE UNDERUTILIZED PLANTS

Nutritional, Antinutritional, and
Nutraceutical Aspects

By

V. R. Mohan, PhD

P. S. Tresina, PhD

A. Doss, PhD

APPLE ACADEMIC PRESS

Apple Academic Press Inc.	Apple Academic Press Inc.
3333 Mistwell Crescent	1265 Goldenrod Circle NE
Oakville, ON L6L 0A2	Palm Bay, Florida 32905
Canada USA	USA

Exclusive worldwide distribution by CRC Press, a member of Taylor & Francis Group

International Standard Book Number-13: 978-1-77188-771-7 (Hardcover)
International Standard Book Number-13: 978-0-42902-302-6 (eBook)

Library and Archives Canada Cataloguing in Publication

Title: Wild edible underutilized plants : nutritional, antinutritional, and nutraceutical aspects / by V.R. Mohan, PhD, P.S. Tresina, PhD, A. Doss, PhD.

Names: Mohan, V. R., 1959- author. | Tresina, P. S., 1987- author. | Doss, A., 1979- author.

Description: Includes bibliographical references and index.

Identifiers: Canadiana (print) 20190114991 | Canadiana (ebook) 20190115009 | ISBN 9781771887717 (hardcover) | ISBN 9780429023026 (ebook)

Subjects: LCSH: Wild plants, Edible.

Classification: LCC QK98.5.A1 .M64 2019 | DDC 581.6/32—dc23

CIP data on file with US Library of Congress

Apple Academic Press also publishes its books in a variety of electronic formats. Some content that appears in print may not be available in electronic format. For information about Apple Academic Press products, visit our website at **www.appleacademicpress.com** and the CRC Press website at **www.crcpress.com**

About the Authors

V. R. Mohan, PhD

Professor, Department of Biomedical Science & Technology, Noorul Islam Centre for Higher Education, Kumaracoil, Thuckalay - 629 180, Kanyakumari District, Tamil Nadu, India.

V. R. Mohan, PhD, is a former Associate Professor and Head of Botany at V. O. Chidambaram College, Tuticorin, India. His research areas are seed biochemistry and ethnopharmacology. He earned his PhD in the field of wild edible legumes and now has nearly 30 years of undergraduate as well as post-graduate teaching experience, along with 24 years of research experience. During his teaching service, he has supervised 41 PhD students on ethnomedicinal plants and 17 MPhil students in the same field. He has published more than 400 research articles related to ethnomedicinal plants in various international and national peer-reviewed refereed journals. He is a resource person in this field. His other works include documentation of ethnomedicinal plants that are endemic to Southern Tamil Nadu. He had also surveyed and documented several ethnomedicinal plants along with their pharmacognostical and pharmacological aspects) in the region of Southern Western Ghats and Tamil Nadu. Dr. Mohan organized an UGC-sponsored seminar and an In-Service Training Program sponsored by Tamil Nadu State Council for Science and Technology. He also serves as a reviewer for various international journals. Furthermore, he has received "PEARL Foundation Best Senior Scientist Award" conferred by PEARL, A Foundation for Educational Excellence. His Google Scholar citation is 3679 with an H index of 30 and an 117 index of 97.

P. S. Tresina, PhD

Assistant Professor, PG and Research Department of Botany,
V.O. Chidambaram College, Tamil Nadu, India

P. S. Tresina, PhD, is an Assistant Professor of Botany at V. O. Chidambaram College, Tuticorin, India. She has five years of teaching experience and nine years of research experience in seed biochemistry as well as ethnopharmacology. During her teaching service, she has supervised 2 MPhil students in the same field. She has published nearly 93 articles in various international and national peer-reviewed refereed journals and four book chapters. She has been awarded a Tamil Nadu Government stipend for a full-time scholarship for the year 2009–2010 and has worked as UGC Major Research Project Fellow in the Postgraduate & Research Department of Botany at V. O. Chidambaram College. Dr. Tresina has received a "Best Young Women Scientist Award in Botany," conferred by PEARL, A Foundation For Educational Excellence. Dr. Tresina earned her PhD specializing in seed biochemistry and earned her undergraduate and postgraduate degrees in botany from St. Mary's College, Tuticorin, India.

A. Doss, PhD

Assistant Professor, PG and Research Department of Botany, V.O. Chidambaram College,
Tamil Nadu, India.

A. Doss, PhD, is an Assistant Professor of Botany at V. O. Chidambaram College, Tamil Nadu, India. He has three years of teaching experience and ten years of research experience in the field of ethnopharmacology. He has received his graduation and post-graduation degrees from St. Joseph's College, Trichy, India, and Doctor of Philosophy from National College, which is affiliated with Bharathidasan University, India. He has published more than 130 research papers in various international and national peer-reviewed refereed journals as well as two books.

Contents

Abbreviations

CVD	cardiovascular disease
DPPH	1,1-diphenyl-2-picrylhydrazyl
EV	*Echis carinatus* venom
FC	flavonoid content
GGT	glutamyl transpeptidase
GI	glycaemic index
HCC	hepatocellular carcinoma
ISFR	India State of Forest Report
LIKO	Like 1 knockout
MEMP	*M. pruriens* plant
MPE	*M. pruriens* seed aqueous extract
NFE	nitrogen free extractives
NO	nitric oxide
NPC1L1	Niemann-Pick C1-like 1
NTFPs	non-timber forest products
ROS	reactive oxygen species
SHBG	sex hormone binding globulin
TDF	total dietary fiber
WEPs	wild edible plants
WT	wild type
YMP	yam mucopolysaccharides

Foreword

Dr. N. Sivaraj
Principal Scientist (Economic Botany & Plant Genetic Resources), National Bureau of Plant Genetic Resources regional Station, Indian Council of Agricultural Research (ICAR), Rajendranagar, Hyderabad – 500 030, Andra Pradesh, India

The Indian Gene center is endowed with an enormous richness of crop diversity and plants, including wild edible plants. The phytogeographical regions of India are of great antiquity and known for the affluence of plant genetic wealth. Over 17,500 species of higher plants are recorded from these regions in India, and about 30% of this flora is endemic. It constitutes our invaluable assets to meet the growing needs of food and nutritional security of our country. Over 1000 plant species fall under the wild edible category and provide supplementary food resources. All these resources of potential value for current and future needs of civilization are increasingly being exploited, and some of them, because of their over-exploitation, are in danger of extinction. The diversity comprising of this valuable wild edible underutilized plant species of Indian origin/naturalized in India is of great potential value and is an Indian heritage.

I am indeed very happy that Dr. Veerabahu Mohan has brought out this authored book, *Wild Edible Underutilized Plants: Nutritional, Antinutritional and Nutraceutical Aspects,* based on the vast experiences he had in the field of plant genetic resources over three decades. The book is well-written, and the materials are well-synthesized in several chapters. It covers different facets of wild edible plant genetic resources activities, namely survey and documentation of tribal edible plants; consumption patterns; nutritional and antinutritional aspects; roots, tubers, rhizome, leafy vegetable, and meristems; wild legumes – L-DOPA content, seed and seed composition: nutritional, proximate composition, mineral composition, soluble carbohydrate, starch, free amino acids, vitamins (niacin and ascorbic acid); nutraceutical aspects and pictorial representations of wild plants in natural condition. These chapters will broaden the outlook of readers to assess wild plants diversity and its importance in national food and nutritional security.

Finally, I express my sincere appreciation and congratulate Dr. Veerabahu Mohan for a commendable work in bringing out a well-written, well-synthesized, and well-illustrated book on wild edible underutilized plants. I am sure that this book will be of immense value to the graduate and postgraduate students, teachers, and researchers. It is hoped that this book will create considerable interest and greater concern in comprehending uses of native edible plants, their diversity, conservation, and potential present and future uses.

—N. Sivaraj, PhD
Principal Scientist
(Economic Botany & Plant Genetic Resources)

Preface

Wild underutilized edible plants were an important source of food for mankind before the dawn of civilization and the domestication of the present day fruits. Cavemen in the forests also depended on these plants and passed on precious information on the efficacy and selection of wild species of plants from generation to generation. Thus, the present day horticulture came into existence. Also, the wild varieties of plants, yielding edibles, growing throughout the country, contributed directly to the cultural heritage of India. A conceptual approach to wild edible plants, including the contribution of such species to people's diets and daily lives, focusing on nutritional and cultural value, food sovereignty and security, as well as the huge legacy for future generations, leads to a general overview of new tendencies and availability of wild plant resources according to geographic regions.

Even today, these wild edible plants are eaten in plenty by the local people, as they are commonly available in abundance in their habitats. The wild trees yielding edible fruits also provide firewood, serve as windbreaks and fences, supply leaf fodder and act as raw material for many useful medicinal preparations. Valuable traits of these wild edible plants, such as resistance to diseases, winter-hardiness, resistance to drought, and possession of superior vigor can be incorporated into their cultivated relatives with a vision of improving them.

In developing countries, a significant hindrance to human survival is the ever-increasing gap between food availability and the rising human populace. Food insecurity results in less consumption of fruits and vegetables and leads to mineral and vitamin deficiency for individuals in these regions. *Wild Edible Underutilized Plants: Nutritional, Antinutritional, and Nutraceutical Aspects* focus on the usage of wild plants in order to reduce food insecurity and malnutrition. Wild plants contain potential bio-molecules both in organic and inorganic combinations. They are a crucial, reasonably priced and beneficial source of vitamins, antioxidants, fiber, minerals and other nutrients for many economical deprived natives. These plants have high nutraceutical value and are used for a wide range of ailments and have the potential to protect the human body from cancer, diabetes, inflammatory, and cardiovascular diseases. As this book exclusively focuses on edible wild plants and their nutritional, anti-nutritional, and nutraceutical aspects, this text is exceptionally valuable to any researcher studying novel potential solutions to food deficiency in the developing world.

Customary and indigenous food resources comprise the bedrock of the diversity in traditional and indigenous food systems of communities in a developing country. The underutilized food resources have a much higher nutrient content than globally known species or varieties commonly produced and consumed. With climate vagueness, there is an imperative need to diversify our food base to a wider range of food crops species for greater system resilience. Traditional and indigenous food crops are less damaging to the environment and address cultural needs; they also preserve the cultural heritage of local communities. Successful food systems in transition effectively draw on locally available food varieties and traditional food culture. It is imperative to collect and document local knowledge, encompassing all aspects of indigenous and underutilized foods, from traditional beliefs to utilization and agronomic practices. Promoting the use of underutilized wild edible species needs to be achieved by highlighting their importance in their current production areas as well as exploiting further opportunities to extend their production and consumption. This information should be useful for both product development and awareness-raising. We sincerely hope that this book will serve as an important step towards conserving and continue utilizing traditional and indigenous food resources.

Unfortunately, the information available on the wild edible plants is rather scanty. Therefore, the scientific and systematic harnessing of these plants has not been possible. In the absence of such information in the form of handbook or manual on wild edible plants, the farmer and orchardist feels greatly handicapped and is reluctant to extend the cultivation of such unusual fruits. Till today, the references to wild edible plants are either available in the botanical floras or in encyclopedic works. These books are not easily available everywhere. Moreover, these are expensive for a common man. Further, these books are difficult to refer and also lack in comprehensive information.

In the present work, the authors have attempted to bring out detailed information on various characteristics of nutritional, anti-nutritional and nutraceutical importance. It is hoped that the book will prove useful to the students and the researchers in horticulture, forestry, and botany. This text focuses on underutilized wild plants that can help to reduce food deficiency in developing nations. Edible wild plants are viewed as a potential solution for overcoming food insecurity for families in these regions, with a specific focus on sustainable production and conservation measures.

The principle aim of this book is to provide comprehensive information about the wild edible plants. We earnestly anticipate that this book will serve as a textbook and a guide to the readers interested in wild edibles in the developing countries. The coverage includes ample depth to prepare researchers to go further in the research work.

—Dr. V. R. Mohan, PhD
Dr. P. S. Tresina, PhD
Dr. A. Doss, PhD

CHAPTER 1

Introduction

Wild edible plants (WEPs) refer to species that are neither cultivated nor domesticated, but are available from their wild natural habitat and used as sources of food. Despite the primary reliance of most agricultural societies on staple crop plants, the tradition of eating WEP products continues in the present day. In addition to their role in closing food gaps during periods of drought or scarcity, WEPs play an important role in maintaining livelihood security for many people in developing countries (Afolayan and Jimoh, 2009). WEPs have been a focus of research for many ethnobotanists in recent decades. Currently, there is a renewed global interest in documenting ethnobotanical information on neglected wild edible food sources (Bharucha and Pretty, 2010). Since traditional knowledge on WEPs is being eroded through acculturation and the loss of plant biodiversity along with indigenous people and their cultural background, promoting research on wild food plants is crucial in order to safeguard this information for future societies (Asfaw, 2009).

More than 12,000 plant species considered edible by humans, e.g., plants for human consumption account for about 5% of the total plant species of the world (Kunkel, 1983). It is estimated that in India about 800 species are consumed as wild edible plants (Singh and Arora, 1978). Wild edible plants (WEPs) refer to species that are harvested or collected from their wild natural habitats and used as food for human consumption (Lulekal et al., 2011; Heywood, 2011; Seal, 2012). WEPs play a major role in meeting the nutritional requirement of the tribal population in remote areas (Sundriyal and Sundriyal, 2001). WEPs serve as supplementary food for non-indigenous people and are one of the primary sources of cash income for poor communities (Uprety et al., 2012; Ghorbani et al., 2012; Menendez-Baceta et al., 2012). WEPs have an important role in ensuring food security and improve nutrition in the diets of many people in developing countries (Lulekal et al., 2011; Ghorbani et al., 2012). WEPs are potential sources of species for domestication and provide valuable genetic traits for developing new crops through breeding and selection (Pandey et al., 2008; Ford-Lloyd et al., 2011). India has a tribal population of 42 million, of which some 60 percent live in forest areas and depend on forests for various edible products (Jana and Chauhan, 1998). Several researchers have documented the WEPs used in the diet by tribes in various parts of India, nutritive values of WEPs and need for a revival of knowledge associated with WEPs (Sundriyal and Sundriyal, 2004; Konsam et al., 2016; Mahapatra and Panda, 2012; Khyade et al., 2009).

World over, the tribal population still stores a vast knowledge on the utilization of local plants as food material and other specific uses (Sundriyal et al., 1998). The tribal communities draw their sustenance mainly from the forests, which provide them food plants and other material requirements. Their lives are much dependent on forest or natural plant wealth. The biological wealth is so intrinsically important to the lifestyle and systems of the indigenous communities that wild plants make an important contribution for the sustenance of local communities. They play a significant role in a wide range of agricultural systems as a source of wild food and fuelwood, and have an important socio-economic role through their use in medicines, dyes, poisons, shelter, fibers and religious and cultural ceremonies (FAO, 1999). Wild edible plants not only supplement the food quantity but also make a significant contribution to the populations' nutrition throughout the year (Griveti and Britta, 2000; Herzog et al.; Britta, 2001; Britta et al., 2001; Britta et al., 2003). Although the principle role of these plants is to supplement the food cultivated in home gardens and other forms of agriculture, many of the species grown or wild-harvested are reported to have both therapeutic and dietary functions (Britta et al., 2003). The sale form the surplus of their collection also adds to their income significantly. Research in several regions has also illustrated that many wild plants that are retained in local food cultures are inseparable from traditional therapeutic systems (Britta et al., 2003; Etkin and Ross, 1982; Fleuret, 1993; Gessler and Hodel, 1997; Moreno-Black et al., 1996). Emphasis on the conservation and management of wild edible plants will help enhance and maintain the religion's biodiversity with little adverse impact on the biodiversity (Britta, 2001). For this purpose identification of the edible species according to the local preferences is necessary (Ambe Guy-Alain and Malaisse Francois, 2001).

The present-day wild edible plants are particularly useful during the famine and similar scarcity situation. Even during normal times, wild plants provide materials of diet to the less advanced section of the human community, often referred as Tribals/Adivasis in India who generally inhabit hilly and other less accessible tracts in both developed and developing countries (Arora and Pandey, 1996). In India, it is estimated that about 800 species are consumed as wild edible plants, chiefly by the tribal people (Singh and Arora, 1978). The present paper reviewed on wild edible plants documented in different parts of India and their utilization by the tribes. The term "wild food" is used to describe all plant resources outside of agriculture areas that are harvested and collected for the purpose of human consumption in forests, Savannah, and other bushland areas. Wild foods are incorporated into the normal livelihood strategies of many rural people, shifting cultivation, continuous croppers or hunter-gatherers (Bell, 1995). Indigenous knowledge of wild edible plants is important for sustaining utilization of those plant species (Jasmine et al., 2007).

Emergency food is often termed as wild food, as apparently, it implies the absence of human interference and management, but in fact, such food plants result from the co-evolutionary relationship between human and environment. Emergency food plants can be divided into two broad groups on the basis of consumption, the one which is not consumed regularly on account of their limited seasonal availability while others are frequently consumed due to easy availability.

The value of wild edible vegetables in food security has not been given sufficient attention in India. Consequently, there are no formal interventions that seek to encourage people to use traditional vegetable as sources of essential nutrients. For many years the importance of wild plants in subsistence agriculture in the developing world as a food supplement and as a means of survival during drought and famine has been overlooked. Nevertheless, whereas the rich indigenous knowledge on the medicinal use of wild plants has been relatively well-documented, research, particularly concerning the socio-economic, cultural, traditional and nutritional aspects of wild food plants still lacks adequate attention. There are at least 3000 edible plant species known to man, with merely 30 crops contributing to more than 90% of the world's calorie intake and only 120 crops are economically important on a national scale (Cooper et al., 1996). There are 1532 edible wild food species in India, mostly from the Western Ghats and Himalayan regions (Arora and Pandey, 1996). Similarly, in Eastern Ghats region also, several tribals are using wild plants as food.

Most of them depend on forest resources for their livelihood due to lack of agriculture land and take edible forms of flowers, roots, fruits, tubers, rhizomes, leaves, etc. for food. Wild food plants are able to fill a variety of plant are in demanding because its availability is more compared to other parts of the plant. Some sporadic work has done on the wild edible plants used by tribal people, but no detailed study about the traditional use of wild plants as food is available in Andhra Pradesh. The paper highlights some of the important wild food plants, which need to be documented for food security in the future.

Use of a large number of wild species by the tribals to meet their diverse requirements is largely due to the prevalence of diversity of vegetation in the area (Katewa, 2003). The use of wild plants is an integral part of their strong traditional and cultural systems and practice that have developed and accumulated over generations. These systems form the basis of local-level decision-making in agriculture, food production, human and animal health and natural resource management (Slikkerveer, 1994).

Consuming wild edibles is part of the food habits of people in many societies and intimately connected to virtually all aspects of their socio-cultural, spiritual life and health (Singh, 2006). It also plays a major role in meeting the nutritional requirement of the tribal population in remote parts of the country throughout year (Setalaphruk and Lisa, 2009; Sundriyal and Sundriyal, 2001; Grivetti and Britta, 2000; Britta, 2001; Britta et al., 2003; Sasi et al., 2011; Hazarika et al., 2012). Wild food plants play a very important role in the livelihoods of rural communities as an integral part of the subsistence strategy of people in many developing countries (Jadhav, 2011). India is one of the second largest human populations on this planet, and 75% of the population is living inthe rural areas. Most rural communities depend onthe wild resources including wild edible plants to meet their food needs in periods of food crises, as well as for additional food supplements (Rashid, 2008). It is estimated that in India about 800 species are consumed as food plants, chiefly by the tribal inhabitants (Singh and Arora, 1978). Wild plants have since ancient times, played a very important role in human life; they have been used for food, medicines, fiber, and other purposes and also as fodder for domestic animals. In the search for wild edible food plants many of which are potentially valuable for a human being has been identified to maintain a balance between population growth and agricultural productivity, particularly in the developing countries (Kanchan, 2011).

Various publications provided detailed knowledge about the utilization of wild plants as food in a specific location around the world. Studied conducted in Africa by Zemede showed that wild plants are essential components of many African diets, especially in the period of seasonal food shortage (Zemede, 1997). A study conducted by Wilson in Zimbabwe revealed that some poor household relies on wild fruits as an alternative to cultivated for a quarter of all dry seasons meals (Wilson, 1990). Paster and Gustavo in their study conducted on wild edibles found that 57 wild edible plants species are consumed, in 118 different ways as a source of food by charlotte people of Argentina (Pastor and Gastavo, 2007). Francesca and Francesca described the importance of 188 wild food plant species used popularly in the Sicily (2007). Javier compiled and evaluated the ethnobotanical data available on the wild edible plants traditionally used for human consumption in Spain. A total of 419 wild plant species belonging to 67 families were discussed with respect to the part used, localization, method of consumption, and harvesting time. This study showed that the reported wild edibles are the essential components of many Spanish diets especially during various traditional events and fairs (Javier et al., 2006). Victoria described the cultural, practical and economic value of wild plants by applying a quantitative technique in the Bolivian Amazon and concluded that wild plants play an important role in the daily life of local inhabitants (Victoria et al., 2006). A study conducted by Athena on Paphos and Larnaca countryside of Cyprus revealed that inhabitants of these areas subsisted primarily on pastoralism and agriculture and therefore preserves the traditional knowledge on wild edible plants (Athena et al., 2006). Ana and Mariana studied the pattern use and knowledge of wild edible plants in distinct ecological environments, from Northwestern Patagonia and found that knowledge and consumption of wild edible plant follow a pattern according to ecological conditions of gathering environments, as well as the cultural heritage of the Paineo people (Ana et al., 2006). Agarahar Murugkar and Subbulakshmi studied the nutritive value of wild edible fruits, berries, nuts, roots, and spices consumed by the Khasi tribes of India. They concluded that the wild plants eaten by Khasi tribe are a good source of nutrients and considering their low cost and easy availability, need to be popularized and recommended for commercial exploitation

(Murugkar and Subbulakshmi, 2005). Maikhuri studied the nutritional value of some lesser-known wild food plants and their role in tribal nutrition (Maikhuri, 1991). Rawat reported some common wild fruits of Garhwal Himalaya (Rawat et al., 1994). Rakesh found that wild edibles are playing an important role in rural development in the central Himalaya Mountains of India (2004). A study conducted by Debarata on the wild food plants of Midnapore, West Bengal showed that 31 wild edible plant species are frequently consumed during the flood and droughts (2004).

However, the century-old traditional knowledge system for the utilization of wild plants is depleting very quickly (Hamilton, 1995; Kiremire, 2001). Modern scientific researchers are also trying to value these traditional food items to fill the gaps between the growing population and food production. These natural products are coming from wild sources and their herbal properties unknowingly flow in diverse ethnic preparations. Such preparation must be variable with the local availability plant resources, forest types, geographical area, and more specifically by different culture and tradition of ethnic groups in Northeast India. Survey and documentation of wild edible plants and their utilization for food have been conducted in several parts of the country (Sasi and Rajendran, 2012; Pfoze et al., 2012).

A major objective of an ethnobotanical investigation into wild food plants is the documentation of indigenous knowledge associated with these plants. Comparative studies on WEPs in different cultures or ethnic groups of a country or among different countries may contribute to the identification of the most widely used species for further nutritional analysis (Termote et al., 2009; De Caluwé, 2010a and 2010b). Nutritional analysis results provide clues to aid the promotion of those species that have the best nutritional values which help to ensure dietetic diversity and combat food insecurity (Tardio et al., 2006).

A considerable amount of research has been conducted worldwide on WEP ethnobotany with an emphasis on field surveys and documentation, to cite but a few: Asfaw and Tadesse (2001); Pieroni et al. (2002); Ertug (2004); Reyes-Garcia et al. (2005); Balemie and Kibebew (2006); Tardio et al. (2006); Arenas and Scarpa (2007); Rashid et al. (2008); Asfaw (2009); Giday et al. (2009); and Teklehaymanot and Giday (2010). Moreover, research on nutritional value and health benefits of WEPs has been reported from Grivetti and Ogle (2000); Ohiokpehal (2003); Heinrich et al. (2005); Balemie and Kibebew (2006); Termote et al. (2009, 2010 and

2011), De Caluwé (2010a and 2010b); Beluhan and Ranogajec (2010) and Feyssa et al. (2011). Regardless of the numerous efforts to document WEPs and associated indigenous knowledge, underestimation of the value of these WEPs can lead to the neglect of ecosystems that nurture them and the indigenous knowledge systems that are related to them (Pilgrim et al., 2008).

There are at least 3000 edible plant species known to mankind, but just about 30 crops alone contribute to more than 90% of the world's calorie intake, and only 120 crops are economically important at the national scale (FAO, 1993). This shows that several hundreds of species remain discarded or unnoticed at the hands of various human societies. Among the edible plant diversity, many are nutritionally or otherwise important. For example, Quinoa (*Chenopodium quinoa*), a staple grain of *Incas* is little known to the modern world, yet it is one of the world's most productive sources of protein. Similarly, a number of such little known crops and edible species found in the wild are not getting recognition, though they play a crucial role in the food security of tribal and rural families. For instance, various wild species of *Dioscorea*, *Colocasia*, and *Amaranthus*, which are the source of vitamins and nutrients, supplement the food needs of a multitude of families who live near to forests (Roy et al., 1998). Wild food also contributes to the household income security of millions of forest-dependent communities. In India, those who collect species such as gooseberry, garcinia, and honey for the market are mostly dependent on it as their major source of income (Muralidharan et al., 1997).

Wild edible fruits play a significant role in rural areas by providing nutrient supplementary diet and generating side income to the poor people. Wild fruits can be considered as rich sources of various vitamins, minerals, fibers and polyphenols which provide health benefits (Narzary et al., 2013; Agbemafle et al., 2012; Prasad et al., 2010; Egbung et al., 2013; Ogbonna et al., 2013; Mahadkar et al., 2013; Amadi et al., 2013; Deshmukn and Rathod, 2013). Consumption of wild fruits reduces the risk of several diseases like diabetes, cancer, coronary heart disease, and neurodegenerative ailment (Ileboye et al., 2013; Rajurkar and Gaikwad, 2012; Saikia and Deke, 2013; Rapheall and Adebayo, 2011; Oko et al., 2012; Sarma et al., 2013; Gireesha and Raju, 2013). A scientific investigation of wild edible fruits is urgently needed to assess the potentiality which would be cultivated and utilized as a source of food material for an ever increasing population.

Human race always depends on nature either directly or indirectly for food, clothes, shelter, and medicine. Nowadays due to the improvement of knowledge and technology, we obtain our food through agricultural practices. However, if we observe the initial days of civilization and evolution of agriculture, all the food plants were discovered from their natural resources from time to time. Still, there are large numbers of plant species, which can be used to fulfill the nutrition requirement of the growing population of the world. Tribals are the part of nature, and they fulfill most of their needs from wild resources. They got knowledge of wild edible plants traditionally. This traditional knowledge is useful to develop new food sources. Exploration of natural resources and documentation of traditional knowledge is necessary.

The nutritional value of many forest foods is not known but appears to be enough information to indicate that forest foods are nutritionally valuable. The studies on the nutritional value of forest food are extremely important as it will encourage people to consume a greater quantity of food and provides them with a better balance of nutrients (FAO, 1989). Wild edibles are important Non-Timber Forest Products (NTFPs) for tribes. According to the India State of Forest Report (ISFR) 2015, the total forest and tree cover is 79.42 million hectares, which is 24.16 percent of the total geographical area. Tribal population of India is 8.6 as per 2011 census. In India, the tribal people depend on forests for their livelihood. The tribal people are very close to nature and have hereditary traditional knowledge of consuming wild plants and plant parts viz. tuber, shoots, leaves, fruits, etc., as a source of food. Although these wild edible plants play an important role in food security, they are ignored. Various tribal sects of India are repositories of rich knowledge on various uses of plant genetic resources (Khoshoo, 1991). Wild edible plants play a major role in meeting the nutritional requirement of the tribal population. Among the various kinds of plants, food plants received the earliest attention of mankind and reflect man's search for knowing more and more about the nutrient qualities of food plants. The primitive man through trial and error has selected many wild edible plants, which are edible and subsequently domesticated them. Modern man neither domesticated the leftover nor has he identified any new food plants in recent times, which are widely accepted; they have improved only a few crop plants.

Williams (1993) emphasized the importance of preserving the new plant resources to broaden the biological diversity in human nutrition.

The importance of wild plant species is being recognized as they provide minerals, fiber, vitamins, and essential fatty acids and enhance taste and color in diets. They can also be used to prevent chronic diseases (cardio-vascular disease, diabetes) in the general population, as well as diseases due to undernutrition like anemia (Williams, 1993; Green, 1993).

It has been noticed that the tribes who still live in their undisturbed forest areas and having the traditional food habit like consumption of a large variety of seasonal foods are found to be healthy and free from most of the diseases (Anonymous, 1995). According to one report from Govt. of India, food deficiency usually prevails in underdeveloped tribal areas (Arora and Pandey, 1996). Still, such tribal groups sustain successfully under adverse conditions as they stick on the alternative source of food, in the absence of wheat and rice and other kinds of conventional staple food plants. A large number of plant species as supplementary food used by tribes of India are reported in 'Dictionary of Indian Folk Medicine and Ethnobotany.' Broad nutrient categories include carbohydrates, fats, proteins, vitamins and minerals such as sodium, calcium, potassium, etc. which are required in comparatively larger amount by the body and therefore called as macro-elements, whereas elements required in smaller amount are called trace or minor elements, for example, iron, zinc, copper, etc.

It is a matter of great pride that among the 18 hot spots known for rich flora in the world over, two are located in India. They are the Eastern Himalayas and the Western Ghats (Khoshoo, 1996). The hill chain of Western Ghats recognized as a region of the high level of biodiversity is under the threat of rapid loss of genetic resources (Gadgil, 1996). The biodiverse nature of the Eastern Ghats is meager. Keeping these facts intact in mind, in the present study, the wild edible food plants have been surveyed and documented from the natural strands of the southeastern slopes of the Western Ghats, Tamil Nadu, India. The Palliyars, Kanikkars, Valaiyans, and Pulayans are the dominant tribal group; inhabit the locality of the study area. The present study focuses on the dependence of the above-mentioned tribals on edible plants and attempts at an exhaustive analysis of the nutritional qualities of such edible plants.

CHAPTER 2

Study Area

The present investigation studies the distribution of the tribals Palliyars, Kanikkars, Valaiyans and Pulayans along the target area, and the ecological conditions, consumption pattern of different food plant and their nutritional potential.

2.1 WILD EDIBLE PLANTS AND THE PALLIYARS: AN INVESTIGATION

The Palliyars are distributed along the Southeastern slopes of the Western Ghats, Tamil Nadu (adjoining Virudhunagar and parts of Madurai and Tirunelveli districts – Area Map).

The present study focuses on the wild edible food plants consumed by the Palliyars settled in the reserve forest area of the Grizzled Giant Squirrel Wildlife Sanctuary. This area lies mostly in Virudhunagar and parts of Madurai district. This wildlife sanctuary is established in the year 1989. It encompasses an area of 480 sq.km. The study also covers tribal hamlets around Puliangudi, Tirunelveli district.

The area of investigation lies between 70° 3′E and 77° 9′E longitude and 9° 1′N and 9° 8′N latitude. This altitude varies from 100 m to 2210 m (MSL). It receives rainfall from both Southwest and the Northeast monsoons. The varied climatic and topographic conditions prevailing in the Sanctuary present remarkable diversity of both flora and fauna.

Variations in the altitude and rainfall always have a bearing on the vegetation in general. The study area consists of tropical evergreen forests, semi-evergreen forests, dry teak forests, southern mixed deciduous forests and dry grassland (Plate Ia).

In the study area the Palliyars, live in several isolated pockets or in small hamlets. Their habitations are known by the following names:

1. Saduragiri	9. Shenbagathoppu
2. Thanniparai	10. Ayyanarkoil
3. Vandipannai	11. Thallaianai
4. Nellikka koottam	12. Shelimbuthoppu
5. Valliammal Nagar	13. Kulirattu
6. Saraupannai	14. Manapadai
7. Petchikeni Koottam	15. Rakkamal Koil Parai
8. Athikoil	16. Kottamalai

2.2 WILD EDIBLE PLANTS OF KANNIKARS: AN INVESTIGATION

The Agasthiarmalai Biosphere Reserve lies between 8° 22′ and 8° 53′ North latitude and between 77° 10′ and 77° 35′ East longitudes in Tirunelveli and Kanyakumari district of Tamil Nadu in the Southern Western Ghats of India (Area map). The boundaries are Ambasamudram and Tenkasi taluks of Tirunelveli districts in the north, Ambasamudram and Nanguneri taluks of Tirunelveli districts in the east, Kanyakumari district in the south and Kerala State in the west.

Geographically, it is a part of the Southwestern tip of the Western Ghats, a region that is known for its species richness, diversity and a high degree of endemism. The Agasthiarmalai Biosphere Reserve area has been recognized as one of the 'hot-spots' for Biodiversity conservation by the IUCN (Ayyanar and Ignacimuthu, 2005). The altitude ranges from 100 m to 1867 m (MSL). It receives rainfall during the South West as well as North East monsoon. The important peaks of the reserve include Agasthiarmalai, Ainthilaipothigai, and Nagapothigai. Tamiraparani, the perennial river of Tamil Nadu originates from Agasthiarmalai (Pothigaimalai) and flows through this sanctuary.

The varied climatic and topographic conditions prevailing in the sanctuary present a remarkable diversity of both flora and fauna. The study area consists of tropical wet evergreen forests, semi-evergreen forests, moist deciduous forests, tropical riparian fringe forests, dry teak forests, dry deciduous forests, umbrella thorn forest, wet temperate forests, and high and low altitude grasslands.

In the study area (Plate Ib), the Kanikkars live either in isolated pockets or small hamlets. Their habitations are known by the following names:

(i) Ingikuzhi,
(ii) Chinnamylar,
(iii) Periyamylar,
(iv) Agastiyarkanikudiyiruppu,
(v) Tharuvattampari Kani Kudiyiruppu (Servalar)
(vi) Kouthalai, and
(vii) Valayar.

2.3 WILD EDIBLE PLANTS OF VALAIYANS: AN INVESTIGATION

The present investigation deals with the distribution of the *Valaiyans* and their dependence on different wild food plants. The *Valaiyans* are distributed in various places of Tamil Nadu in small settlements especially along the hill tracts in Madurai and Tanjore districts (Sampathkumar, 1991). In the census report 1991, the *Valaiyans* are described as a "Shikkari (hunting) caste" in Madurai and Tanjore. In the Madras Gazette, the *Valaiyans* are included in the list of the denotified community (number 220) inhabiting Madurai, Dindigul, Tiruchirapalli, Pudukkottai, Erode and Coimbatore districts. In the manual of Madurai, the *Valaiyans* are referred to as "a low and debased class" (Thurston and Rengachari, 1987). The name *Valaiyans is* supposed to be derived from *Valai* a net, and it has been given to them because they are constantly employed in netting game in the jungles. The present study focuses on the wild edible food plants consumed by the *Valaiyans* settled in the forest areas. The study area lies in Madurai district (Plate II c).

The area of investigation lies between 78°E and 78.02′E longitudes and 100° and 100.2′N latitude. The altitude varies from 2000′–3000′. It receives rainfall from both Southwest and the Northeast monsoons. Variations in the altitude and rainfall always have a bearing on the vegetation in general. The study area consists of tropical evergreen forests, semi-evergreen forests, mixed deciduous forests, and dry grassland.

In the study area, the *Valaiyans* live in several isolated pockets or in small hamlets. Their habitations are known by the following names: Chathiravellalapatti, Kondaiampatti, Muduvarpatti, Saranthangi, Katchaikatti, Mettupatti, Ramakoundanpatti, and Valaiyapatti.

2.4 WILD EDIBLE PLANTS AND THE PALLIYARS AND PULAYANS: AN INVESTIGATION

The objective of this investigation was to assess the richness of wild edible plants utilized by the tribes Palliyars and Pulayans in the Palani Hills. The study area, Palani Hills is situated between latitude 10°05 and 10°30 N and longitude 77°15 and 78°00 (Plate II d). The Palani Hills an eastern spur of the Western Ghats are located in central western Tamil Nadu. Spread over 2,068 sq.km, the Palani Hills are contiguous with the high range Anamalai and Cardamom Hills and form an imposing range of mountains in Dindigul district, Tamil Nadu.

Like other mountain ranges such as the Nilgiris in the southern part of the Western Ghats, the Palani Hills are made up of pre-Cambrian gneisses charnockites and schists, making them one of the oldest mountain ranges in India. In the southwest the Palani Hills rise abruptly from the plains to form an elevated plateau around 1800–2500 m high, its eastern half is composed of hills 1000–1500 m high. The lower Palani Hills rise to an altitude of 800–1500 m while the upper Palani Hills rise to an altitude of 1800–2500 m.

The Palani Hills is one of the internationally recognized 'hot spot' known for its richness and uniqueness of plant wealth. Primarily based on the difference in the altitudes, the flora of this region is broadly classified into the scrub forest (500 to 750 m), deciduous forest (750 to 1000 m), dry evergreen forest (1000 to 1500 m) and moist evergreen forest (1500 to 2000 m). The temperature ranges from 18°C to 32°C, and the mean annual rainfall is 1200 mm of which the highest rainfall is during August to September.

It is believed that the Palliyars are indigenous people of Palani Hills. In the Palani Hills, they are found at an altitude of up to 2200 m. Generally, the Palliyars are illiterate, and they speak Tamil (mother tongue of Tamil Nadu). The Palliyars are also found in the hilly regions of Madurai, Dindigul, Theni, Tirunelveli and Virudhunagar districts. Reportedly the descendants of Palliyars still live in Kukul cave in Kodaikanal set high in the Palani Hills at 2100 m.

Plate - I
Area Map

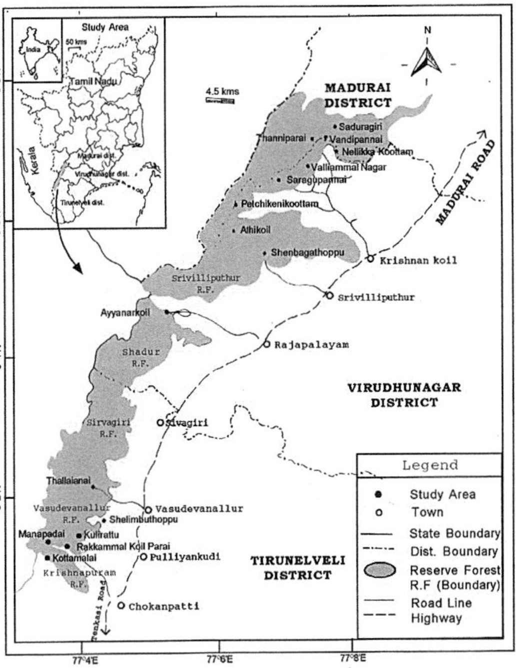

a. Grizzled Giant Squirrel Wildlife Sanctuary

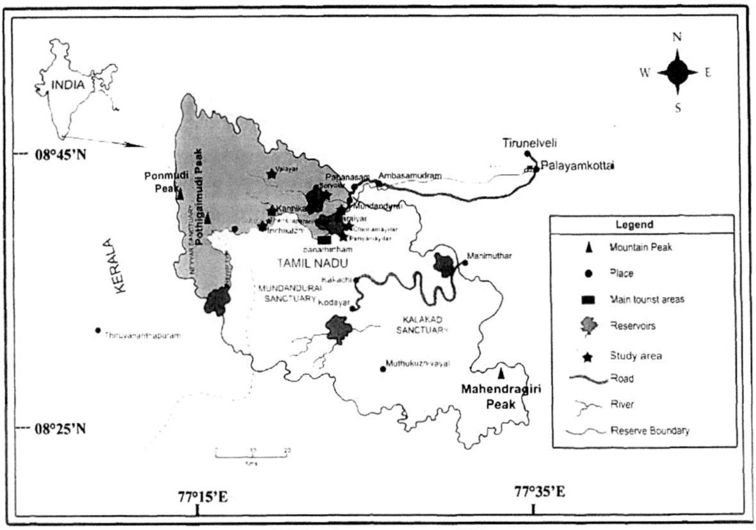

b. Agasthiarmalai Biosphere Reserve

(See color insert.)

Plate - II

Area Map

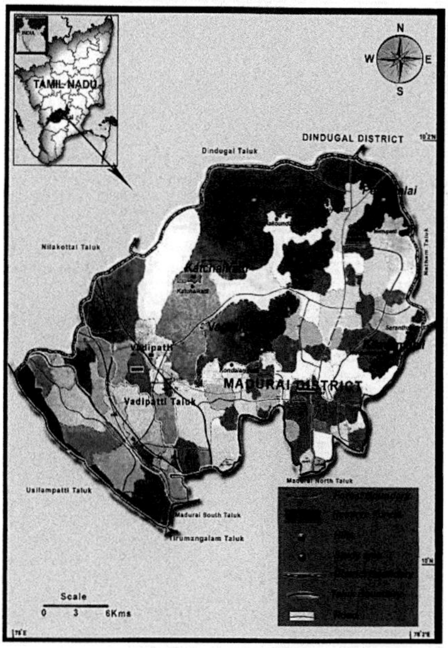

c. Madurai District

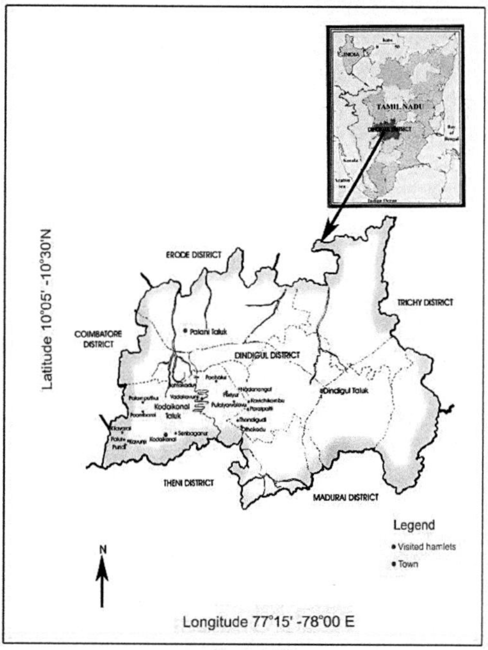

d. Palani Hills

(See color insert.)

CHAPTER 3

Life Style of the Tribals

Tribals are a distinct ethnic group who are usually confined to definite geographical areas. They speak a common dialect and are culturally homogenous. The tribal people of India live mostly in the forests, hills, plateaus, and regions of natural isolation. They are differently named as Adivasi (original settlers), Adim Niwasi (oldest ethnological sector of the populations), Aboriginal (indigenous), Vanavasi (forest inhabitants) and several such other names symbolizing either their ecological, economic or historical or cultural characteristics (Jain, 1987).

The Indian sub-continent is divided into three major tribal zones on the basis of the tribal demography and geography of India: (i) the North and North-Eastern Zone, (ii) the central or middle zone, and (iii) the Sothern zone. The southern tribals are historically the most ancient tribes.

3.1 PALLIYARS

There are as many as 36 types of scheduled tribes in Tamil Nadu. In the serialized list notified by the Government of Tamil Nadu, number 32 denotes the Palliyars. The Palliyars live in the low altitude of Western Ghats. They live in large number in Virudhunagar district, parts of Madurai, Palani Hills and Tirunelveli districts. Of the total population of the scheduled tribes in Tamil Nadu, 5,74,194 the Palliyar tribe accounts for 1,890 which ranks 20th among the total tribal population. At present, the population of Palliyars in the study area is 1,070. Of this, 373 are men, 342 are women, and 355 being children (Authors investigation). The Palliyars live as individual families. They are short, dark complexioned, curly haired with thick protruding lips and a blunt nose with wide nostrils (Plate III a & b).

Generally, a hamlet has about 20 huts. Each hut is thatched with the fronds of Tharagu pull (*Cymbopogon polyneuros*) or lemongrass

(*Cymbopogon citratus*) or leaves of Iechamaram (*Phoenix pusilla*) or coconut leaves (*Cocos nucifera*). They use sand and stone for wall construction. They decorate their houses with the same leaves. They sleep on mats woven with these leaves. They eat tubers, greens, unripe fruits, ripe fruits, etc. they also feed on wild animals and birds like rabbits, rats, deer, hens, etc. They roast besides tubers, the flesh of these animals and birds and eat.

The Palliyars have strong faith in religious customs and practices. They worship "Forest Goddess" and "Goddess Poomadevi" (Goddess Earth). On special days the stay in the 'temples' in groups for two days and worship by offering goats as a sacrifice. On those daysa, large quantities of Mullvallikizhlangu (Dioscorea pentaphylla var. pentaphylla) are collected and offered to the Goddess Earth, 'Poomadevi.'

The Palliyars are monogamous. Elopement is the favorite form of marriage. The elders search for the eloped couples; they bring them back and get them married. Widow remarriage is common. The dead are buried. On the eighth day after death, they perform the last rites. The Palliyars believe in witchcraft. They entertain many curious superstitious beliefs.

The Palliyars are adroit in collecting wild honey. They collect it from the branches of towering tall trees and rock caves skillfully using special techniques. They also collect Kungillium (resin) from the barks of *Canarium strictum*.

They use many herbs to treat diseases. Vaithiyar is a tribal person who is an expert in administering herbal medicines. He collects medicinal plants from various remote parts of the hills. People respect the Vaithiyar and hold in good esteem. The government of Tamil Nadu has established a medicinal plant conservation area in the forest for preservation and development of herbal wealth. It is maintained and conserved with the assistance of the Palliyars.

At present, a free residential school is run by the Government of Tamil Nadu at Maharasapuram for the exclusive educational development of the tribal children.

3.2 KANIKKARS

The Kanikkars belong to the Southern tribal zone. The Kanikkars are also known as Kanikaran or Kani. The Southern tribals are historically more ancient tribes. There are as many as 36 types of scheduled tribes in Tamil Nadu. In the serialized list notified by the Government of Tamil Nadu, the Kanikkars are placed at 7th position. They live in low altitude regions of Western Ghats and live in large numbers. Kanikkars means "hereditary proprietor of the land" thus recognizing their ancient rights over the forest lands.

The Kanikkars are generally very short in stature and meager in appearance (Plate III c & d), from their active habits and scanty food. Some have markedly Negroid features. They are simple and straightforward tribals. They are traditionally a nomadic community. They speak in their own dialect, Kanikkar Bhasha or Malampashi, which is close to the Dravidian language Malayalam. They were once lords of the forest and practiced migratory cultivation but Kanikkars have now to a large extent abandoned this kind of migratory cultivation because of the forest may not be set fire to or trees felled at the unrestricted pleasure of individuals.

Most of the *Kanikkar* tribals have a general knowledge of medicinal plants that are used for first aid remedies, to treat cough, cold, fever, headache, poisonous bites and some other simple ailments. Kanis still supplement their food by gathering roots and tubers from the nearby forest areas. They eat tubers like *Manihot esculenta* and *Dioscorea oppositifolia,* etc. They are extremely hard working and can survive without the help of modern facilities. They are socio-economically backward, and most of them are very poor. They are also engaged in the seasonal collection of honey, bee wax and some minor forest produce. They cultivate edible plants like tapioca, banana, millets and cash crops such as pepper, areca nut, and cashew nut.

3.3 VALAIYANS

In most of the settlements, the Valaiyans live in small huts. They use sand and stone for wall construction. The roof is thatched with Grasses (*Cymbopogon martini* Wats.) or Coconut leaves (*Cocos nucifera* L.) or leaves of palm (*Borassus flabellifer* L.). Most of the huts are gloomy and dark with no windows or any other apertures to admit light or air.

The Valaiyans live in small groups, and each group has a headman referred to as 'Poosari/Kambiliyan' (Plate III e & f). The language of Valaiyans is Tamil. They are Hindus. Most of them are illiterate, and a few of them may read and write.

3.3.1 RITUALS AND RELIGIOUS CEREMONIES

The Valaiyans have strong faith in religious customs and practices. The goddesses of Valaiyans are 'Singa Pidari' (Aiyanar), Padinettampadi Karuppan, Murugan, Pillaiyar, Pandi, Kali, and Pathirakali Amman. Once a year in the day, after new moon in the month of Masi (February to March) they stay in temples for two days and worship offering goats, hares, wild pigs, and other animals as sacrifice. Surrounding the temple Valaiyans raise sacred groves, where the plants are not disturbed in any way and preserved. The floristic components of the sacred groves are: *Azadirachta indica* A. Juss (Meliaceae/Vembu), *Aegle marmelos* (L.) Correa (Rutaceae/Vilvam), *Mangifera indica* L. (Anacardiaceae/Maa).

3.3.2 CEREMONIES RELATED TO PUBERTY, MARRIAGE, AND DEATH

The puberty customs of Valaiyans are interesting. The girl who attained puberty is allowed to stay in a newly constructed seclusion shed for about 15 days. During these days no male is allowed to go to the shed. The Valaiyans observe Kulam and Kothram in marriage. Adult marriage is the rule, and the consent of the maternal uncle is necessary. Remarriage of widows, divorced women are freely permitted. At the marriage ceremony the bridegroom's sister takes up the thali (marriage badge) and after showing it to those assembled ties, it tightly around the neck of the bride. The dead are buried. The final death ceremonies (karmandiram) are performed on the sixteenth day.

They use many wild plants to treat diseases. Vaidhyar is the person who is an expert in administering herbal medicines. He collects medicinal plants from various remote parts of the hills. People respect the Vaidhyar and hold him in good esteem.

3.4 PULAYANS

The Pulayans were apparently the earliest inhabitants of the Palani Hills. The word 'Pulaya' means 'polluted man' and expresses the idea of caste impurity. They eat beef, pork and even rats (Krishna Iyer, 1939). The

Pulayans differ from Palliyars in their eating habits as they eat beef. Because of this habit, the Pulayans are considered inferior by the Palliyars. The Pulayans of the Palani Hills are said to have migrated from Madurai region and seems to be one of the early settlers of the Palani Hills. The Pulayans are the simple folk who lead simple lives in his jungles. A large number of them are plantation workers.

The Tamil speaking Pulayans are also referred to as the Mala Pulayans a group categorized as 'scheduled caste' by the state government of Tamil Nadu. But the group has been consistently contesting for their legitimate tribal identity. They live in small hamlets, in huts and government constructed colonies. The actual sedentary life started with the construction of group houses by the government of Tamil Nadu in the early 1960s.

The Pulayan community is vertically divided into two sub-divisions called Koora and Kanni. The first is further subdivided into 40 sub-sects, the second one into 7 sub-sects. Each sub-sect is called Kootam, which regulates certain social events. Each Kootam has its own deity, which is common to the entire group and once in a year, the members of the same Kootam assemble to worship the deity. Among the Pulayan community, there is a Nattanmaikaran who commands much respect.

3.4.1 PULAYANS: PEOPLE DESCRIPTION

Both men and women are sturdy and resemble the people of the plain in many respects. As for the nature of their dress, it is very simple, as the men wear dhoti and a towel tied around their head (Plate III g & h). But at present, they wear a good dress because of their earnings out of selling

3.4.2 PULAYANS: LIVELIHOOD AND SUSTENANCE

The Pulaiyans like the majority of the hunting and gathering tribes are heavily dependent upon the collection of edible roots, tubers and hunting small games for their sustenance. Along with this they also collect non-timber forest produce (NTFP) like honey fruits, bark, leaves and other edible items for exchange of food materials with forest contractors and plain people. They also cultivate several species of minor millets in small plots located near their hamlets to meet their subsistence requirements.

Plate - III

Tribals of the Area Surveyed

a. An old Palliyar tribal man of Athikoil Sector of Grizzled Giant Squirrel Wildlife Sanctuary

b. A Palliyar family of Rakkammal Koil Parai

c. A very old Kani Woman

d. A Kanikkar family of Servalar

e. Chithirai, a Valaiyan holding *Ceropegia juncea* Roxb.

f. A Valaiyan's family of Valaiyapatti

g. A Pulayan Woman in Pachalur

h. A Pulayan family in Pachalur

(See color insert.)

CHAPTER 4

Materials and Methods

4.1 SURVEY AND COLLECTION

A survey was carried out for a period of 24 months to assess the richness of wild edible plants consumed by the *Palliyars/Kanikkars/Valliyans* and *Pulayans* in the Southeastern Slopes of Western Ghats. Frequent field trips were undertaken to various hamlets where the tribals exist. Information regarding the wild edible plants was gathered by meeting the tribals during explorative field trips and by gaining a good rapport and winning over their confidence. Most of the information included in this study was gathered from the elderly people who have a very long acquaintance with the usage of plants. The filed notebook delineates in all the usage procedures adopted by them. The information thus gathered was adequately cross-checked for reliability and accuracy by interacting with different groups of the tribals from different settlements to confirm the usage pattern as well as differences, if any, in the mode of consumption.

After eliciting detailed information regarding the wild edible food plants and their useful parts, the collected specimens were brought to the laboratory for identification and chemical analyses. Herbaria for all the collected plant specimens were prepared and deposited in the P.G. and Research Department of Botany, V.O. Chidambaram College, Thoothukudi, Tamil Nadu, India.

4.2 BIOCHEMICAL ANALYSIS

From the surveyed plants, tubers, corms, rhizome, roots, apical meristem, greens, seeds, and seed composition were chosen for the biochemical analysis.

4.2.1 PROXIMATE ANALYSIS

4.2.1.1 DETERMINATION OF MOISTURE CONTENT (JANARDHANAN, 1982)

The moisture content of the plant samples was estimated by taking transversely cut samples at a time in a in a known quantity, and the weight was taken before and after incubation in a hot air oven at 80°C for 24 h, followed by cooling in a desiccator. The loss in weight of the sample was calculated as moisture content, and the average value of triplicate determinations was expressed on a percentage basis.

4.2.1.2 DETERMINATION OF CRUDE PROTEIN CONTENT

4.2.1.2.1 Digestion

The nitrogen content in the powdered plant samples was estimated by the micro-Kjeldahl method (Humphries, 1956). One hundred milligrams powdered plant sample was placed into a Kjeldahl digestion flask. In this 2 mL of 5% (w/v) salicylic acid dissolved in concentrated H_2SO_4 was added and mixed well. After 20 min, 0.3 g of sodium thiosulfate was added and heated gently until fumes disappeared. After cooling the contents of the flask, 60 mg of catalyst (a mixture of 1 g copper sulfate, 8 g potassium sulfate, and 1 g selenium dioxide) followed by 1ml of concentrated H_2SO_4 were added. The contents of the flask were digested until they turned apple green in color. The flask was cooled, and the contents were transferred into a 100 mL volumetric flask and made up to 100 mL with distilled water.

4.2.1.2.2 Distillation

Ten milliliters of the aliquot solutions from the volumetric flask was transferred to Paranas micro-Kjeldahl distillation flask. In this 10 mL of 40% (w/v) NaOH solution along with 2 ml of glass distilled water was added. The contents were heated by a bunsen burner. The liberated ammonia was collected in 2% (w/v) boric acid solution containing a drop of double indicator (83.3 mg of bromocresol green + 16.6 mg of methyl red dissolved in 10 mL of 95% ethanol). The contents were titrated against N/50 sulfuric acid. A blank was run simultaneously using all the reagents, and the value of the blank was deducted from the value of the sample before calculation. One milliliter of N/50 H_2SO_4 corresponds 0.00028 g of nitrogen, which forms the basis for calculation of nitrogen content in the sample. Crude protein was estimated by multiplying the sample percentage nitrogen content by a factor 6.25.

4.2.1.3 DETERMINATION OF ETHER EXTRACT (OR) TOTAL CRUDE LIPID CONTENT

Two grams of air-dried powdered sample was extracted with ether in a Soxhlet apparatus for 16 h, according to the AOAC (2005). The ether was evaporated, and the residue was weighed. The average value of triplicate experiments was expressed as a percentage of ether extract or total crude lipid content on a dry weight basis.

4.2.1.4 DETERMINATION OF TOTAL DIETARY FIBER CONTENT

Total dietary fiber (TDF) was estimated by the non-enzymatic-gravimetric method proposed by Li and Cardozo (1994). To determine the TDF, duplicate 500 mg ground samples were taken in separate 250 mL beakers. To each beaker 25 mL water was added and gently stirred until the samples were thoroughly wetted, (i.e., no clumps were noticed). The beakers were covered with Al foil and allowed to stand 90 min without stirring in an incubator maintained at 37°C. After that, 100 mL 95% ethanol was added to each beaker and allowed to stand for 1 hr at room temperature (25±2°C). The residue was collected under vacuum in a pre-weighed crucible containing filter aid. The residue was washed successively with 20 ml of 78% ethanol, 10 mL of 95% ethanol and 10 mL acetone. The crucible containing the residue was dried ≥2 hr at 105°C and then cooled >2 hr in a desiccator and weighed. One crucible containing residue was used for ash determination at 525°C for 5 hr. The ash-containing crucible was cooled for ≥ 2hr in a desiccator and weighed. The residue from the remaining duplicate crucible was used for crude protein determination by the micro-Kjeldahl method as already mentioned. The TDF was calculated as follows:

$$TDF\% = 100 \times \frac{Wr - [(P + A)/100]Wr}{Ws}$$

where *Wr* is the mg residue, *P* is the % protein in the residue; *A* is the % ash in the residue, and *Ws* is the mg sample.

4.2.1.5 DETERMINATION OF ASH CONTENT (AOAC, 2005)

Two grams of oven-dried powdered sample was weighted into a pre-weighted porcelain crucible. The crucible with the powdered sample was placed in an electric muffle furnace set at 600°C and maintained for 2 h. The contents of the crucible were cooled in desiccators and weighed immediately. The difference in weight of the crucible gives the ash content. The ash content was expressed as a percentage on a dry weight basis.

4.2.1.6 DETERMINATION OF NITROGEN FREE EXTRACTIVES (NFE) OR TOTAL CRUDE CARBOHYDRATE CONTENT (MULLER AND TOBIN, 1980)

Percentage of Nitrogen Free Extractives (NFE) was calculated as given below:

$$\% \text{ NFE} = 100 - (\text{CP\%} + \text{EE\%} + \text{TDF\%} + \text{Ash \%})$$

where, (CP = Crude protein; EE = Ether extract; TDF = Total Dietary Fibre)

4.2.1.7 DETERMINATION OF CALORIFIC VALUE

The sample calorific value (kJ) was estimated by multiplying the percentages of crude protein, crude lipid, and NFE by the recommended factors 16.7, 37.7, and 16.7, respectively, used in legume analysis (Siddhuraju et al., 1996).

4.3 MINERAL ANALYSIS

4.3.1 SAMPLE DIGESTION

Five hundred milligrams of air-dried powdered sample was mixed with 10 mL of conc. HNO_3, 4 ml of 60% perchloric acid and 1 mL of conc. H_2SO_4 and the contents were kept undisturbed overnight.

After that, it was heated on a hot plate containing conc. H_2SO_4 in a beaker until the brown flumes ceased coming out and then allowed to cool. After cooling, it was filtered through Whatman No: 42 filter paper. After that, the filtrate was made up to 100 mL with distilled water.

Minerals like zinc, copper, iron, and manganese were analyzed by employing Atomic Absorption Spectrometer (Issac and Johnson, 1975). The other five minerals (sodium, potassium, calcium, magnesium, and phosphorus) were estimated by following different methods.

4.3.1.1 ESTIMATION OF SODIUM AND POTASSIUM

Sodium and potassium were estimated by using Flame photometer (model Elico). The sodium and potassium contents were calculated by referring to the calibration curves of sodium and potassium, respectively, and expressed as mg 100 g^{-1} of seed flour.

4.3.1.2 ESTIMATION OF CALCIUM AND MAGNESIUM (JACKSON, 1967)

4.3.1.2.1 Calcium

Five milliliters of triple acid digested extract was taken in a China dish. To this 10 mL of 100% (w/v) NaOH and 0.1 g of Murexide indicator powder [40 g of potassium sulfate or potassium chloride was added and ground with 10 g ammonium chloride and 0.2 g of Murexide (ammonium purpurate)]. The solution was then titrated against 0.02 N versenate (19 g of EDTA was dissolved in 5 liters of distilled water) was standardized against 0.2 N Na_2CO_3 solutions and adjusted will the color changed from red to violet.

4.3.1.2.2 Calcium and Magnesium

Five milliliters of triple acid digested extract was taken in a China dish, to this 10 mL of ammonium chloride – ammonium hydroxide buffer (pH 10) and a few drops of Eriochrome black T indicator (0.1 g of Eriochrome black T dissolved in 25 mL of methanol containing 1 g of hydroxylamine hydrochloride) were added and titrated against 0.02 N versenate solution until the color changed from red to blue.

4.3.1.2.3 Calculation

Percentage of calcium in the sample = Titre value of calcium × (100/5) × (100/0.5) × 0.0004

Percentage of magnesium = Titre value of calcium + magnesium – titre value of calcium or titre value of calcium + magnesium × 0.96

Calcium and magnesium contents were expressed as mg 100 g^{-1} of seed flour.

4.3.1.2.4 Estimation of Phosphorus (Dickman and Bray, 1940)

One milliliter of triple acid digested extract was pipetted out into 100 mL volumetric flask. To this 50 mL distilled water was added, followed by 5 mL ammonium molybdate–sulfuric acid reagent (solution A: 25 g of ammonium molybdate dissolved in 100 mL of distilled water. Solution B: 280 mL of conc. H_2SO_4 diluted to 800 mL. Solution A was added slowly with constant stirring to solution B, and the volume was made up to 1000 mL with distilled water). The blue color developed after adding six drops of 25% (W/V) stannous chloride solution. The total volume was made up to 100 mL. The intensity of the blue color was measured at 650 nm in a spectrophotometer. The phosphorus content present in the sample was calculated by referring to a standard graph of phosphorus using potassium dihydrogen phosphate (KH_2PO_4) as standard and expressed as mg 100 g^{-1} of seed flour.

4.3.1.2.5 Estimation of Micronutrients by Atomic Absorption Spectrophotometer (Issac and Johnson, 1975)

By feeding the sample on an Atomic Absorption Spectrophotometer [Elico UV-VIS Spectrophotometer SL 150, ELICO – LTD], the following elements were estimated with appropriate wavelengths.

Name of the mineral	Wavelength used for estimation
Iron	248.3 nm
Copper	324.5 nm
Zinc	213.9 nm
Manganese	279.4 nm

The mineral contents were expressed as mg 100 g^{-1} of seed flour.

4.4 EXTRACTION AND ESTIMATION OF TOTAL SOLUBLE CARBOHYDRATE AND FREE AMINO ACIDS

4.4.1 EXTRACTION

One g air-dried powdered sample was suspended in 1:5 (w/v) hot 80% ethanol and extracted for 15 min. at 75°C. The pellet was re-extracted twice with an equal volume of hot 80% ethanol. The ethanol extracts were pooled, clarified by centrifugation at 5000 rpm for 10 min, the clear supernatant was evaporated to 1–2 mL volume. The content was diluted to 50 mL with distilled water. This content was used as the source material for the estimation of total soluble carbohydrates and total free amino acids.

For greens, one g of air-dried powdered sample was ground with 5 mL of 80% methanol. The extract was centrifuged at 5000 rpm for 15 minutes. The supernatant was collected and made up to 10 mL with 80% methanol. To this, 10 mL of petroleum ether was added and mixed well. The lower layer was taken for carbohydrate estimation.

4.4.2 ESTIMATION OF TOTAL SOLUBLE CARBOHYDRATE (YEMM AND WILLIS, 1954)

From suitable aliquots of the above extract total soluble carbohydrates were estimated by the anthrone reagent method of using glucose as a standard at 620 nm in a spectrophotometer. The average value of triplicate determinations was expressed as g/100 g of powdered samples.

4.4.3 *ESTIMATION OF FREE AMINO ACIDS*

From suitable aliquots of the above extract, the free amino acid content was quantified by a modified ninhydrin method, using leucine as a standard in a spectrophotometer at 540 nm. The average value of triplicate determinations was expressed as g/100 g of powdered samples.

4.5 EXTRACTION AND ESTIMATION OF STARCH (SADASIVAM AND MANICKAM, 1996)

4.5.1 *EXTRACTION*

One g of the air-dried powdered sample was ground with 80% hot ethanol to remove the sugars. The residue obtained after centrifugation was repeatedly washed with 80% hot ethanol and the washed residue was allowed to dry. To the dried residue 5 mL of distilled water and 6.5 mL of 52% perchloric acid were added, and it was allowed to stand for 20 min at 0°C. For the extraction of starch, the supernatants were taken by repeated centrifugation using fresh perchloric acid, and the supernatants were pooled. The final volume was made up to 100 mL, and the supernatant was used as the source material.

4.5.2 *ESTIMATION*

0.05 to 0.1 mL aliquots of the above extract were pipette into different test tubes, and the volume was made up to 1 mL with distilled water. 4 mL of ice-cold anthrone was added into each test tube and heated for 8 min. in a boiling water bath. After cooling the intensity of green color was read at 630 nm in a spectrophotometer against a reagent blank. The starch content present in the samples was calculated by referring to a standard of starch using glucose as standard. The values obtained were multiplied by a factor 0.9 to quantify the starch content. The average value of triplicate determinations was expressed as g/100 g of powdered samples.

4.6 EXTRACTION AND ESTIMATION OF ASCORBIC ACID (SADASIVAM AND MANICKAM, 1996)

4.6.1 EXTRACTION

Three gram of air-dried powdered sample was ground with 25 mL of 4% oxalic acid and filtered. 10 mL of the filtrate was taken in a conical flask. Bromine water was added drop by drop with constant stirring to remove the enolic hydrogen atoms in the ascorbic acid filtrate (till the filtrate turned orange-yellow). The excess of bromine was expelled by blowing in the air. The final volume was made up to 25 mL with 4% oxalic acid solution, and it was used as the source material.

4.6.2 ESTIMATION

Two mL aliquots of the above extract were pipetted into each of the different test tubes, and the volume was made up to 3 mL with distilled water. One mL of 2% DNPH (2,4-dinitrophenyl hydrazine) reagent and 1 or 2 drops of 10% thiourea were added to each test tube. The contents of the test tubes were mixed thoroughly and incubated at 37°C for 3 hours. After incubation, 7 mL of 80% sulfuric acid was added to each test tube to dissolve the orange-red osazone crystals, and the absorbance was measured at 540 nm against a reagent blank. The ascorbic acid content present in the sample was calculated by referring to a standard graph of ascorbic acid and expressed as mg 100 g^{-1} of powdered samples.

4.7 EXTRACTION AND ESTIMATION OF NIACIN (SADASIVAM AND MANICKAM, 1996)

4.7.1 EXTRACTION

Five gram of air-dried powdered sample was extracted with 30 mL of 4N sulfuric acid and steamed for 30 minutes. It was allowed to cool and made up to 50 mL with distilled water. This suspension was filtered through Whatman No: 1 filter paper to 25 mL of the filtrate, 5 mL of 60% basic lead acetate was added. The pH of the above suspension was adjusted to 9.5 and centrifuged. 2 mL of conc. H_2SO_4 was added to the supernatant. The supernatant with conc. H_2SO_4 was allowed to stand for 1 h and centrifuged again. Five mL of 40% $ZnSO_4$ was added to the supernatant, and the pH was adjusted to 8.4. The supernatant was collected by centrifugation, and the pH was adjusted to 7. This supernatant was used as the source material.

4.7.2 ESTIMATION

One mL aliquot of the above extract was pipetted into different test tubes, and the volume was made up to 6 mL with distilled water. Three mL of cyanogen bromide was added, and the contents of the tubes were shaken. One mL of 4% aniline was added to each test tube after 10 minutes. The yellow color, which developed after 5 min, was read at 420 nm against reagent blank. The niacin content present in the samples was calculated by referring to a standard graph of niacin using as standard and expressed as mg 100 g^{-1} of powdered samples.

4.8 ANTINUTRITIONAL FACTORS OF THE PLANT SAMPLES

4.8.1 EXTRACTION AND ESTIMATION OF TOTAL FREE PHENOLICS AND TANNINS

4.8.1.1 EXTRACTION (MAXON AND ROONEY, 1972)

One gram of air-dried plant samples flour was taken in a 100 mL flask, to which added 50 mL of 1% (v/v) HCl in methanol. The samples were shaken in a reciprocating shaker for 24 h at room temperature. The contents were centrifuged [REMI Cooling Centrifuge, Model C-24, Mumbai, India] at 10,000 g for 5 min. The supernatant was collected separately and used for further analysis.

4.8.1.2 ESTIMATION OF TOTAL FREE PHENOLICS (SADASIVAM AND MANICKAM, 1996)

One milliliter aliquots of the above extract were pipetted out into different test tubes to which 1 mL of folin-ciocalteu's reagent followed by 2 mL of 20% (w/v) Na_2CO_3 solution was added and the tubes were shaken and placed in a boiling water bath for exactly 1 min. The test tubes were cooled under running tap water. The resulting blue solution was diluted to 25 mL with distilled water, and the absorbance was measured at 650 nm with the help of a spectrophotometer [Elico UV-VIS Spectrophotometer SL 150, ELICO – LTD]. If precipitation occurred, it was removed by centrifugation at 5000 g for 10 min before measuring the absorbance. The amount of phenolics present in the sample was determined from a standard curve prepared with catechol. A blank containing all the reagents minus plant extract was used to adjust the absorbance to zero. Average value of triplicate estimations was expressed as g/100 g of the seed flour on a dry weight basis.

4.8.2 ESTIMATION OF TANNINS

From suitable aliquots of the above extract, tannin content was quantified by Vanillin-HCl method of Burns (1971) using phloroglucinol as a standard at 500nm with a spectrophotometer [Elico UV-VIS Spectrophotometer SL 150, ELICO – LTD]. The average values of triplicate estimations of all samples were expressed as g 100 g^{-1} seed flour on a dry weight basis.

4.8.3 EXTRACTION AND ESTIMATION OF L-DOPA (3, 4-DIHYDROXYPHENYLALANINE) (BRAIN, 1976)

One gram of air-dried was extracted with 5 mL of 0.1 N HCl over a boiling water bath for 5 min. After cooling, an equal volume of ethanol was added. The mixture was shaken mechanically for 10min. The contents were centrifuged [REMI Cooling Centrifuge, Model C-24, Mumbai, India] at 5000 rpm for 10 min. The supernatant was retained. The pellet was re-extracted with an equal volume of ethanol, and the extract was clarified by centrifugation. Both the supernatants were combined and made up to a known volume with ethanol.

L-DOPA content in the extract was quantified by measuring the Ultra-Violet light absorption at 283 nm in a spectrophotometer [Elico UV-VIS Spectrophotometer SL 150, ELICO – LTD] after correction for background absorption using L-DOPA (Sigma Chemical) as a standard. The contents of L-DOPA in seed flour of all the samples were calculated and expressed as a percentage on a dry weight basis.

4.8.4 EXTRACTION AND ESTIMATION OF HYDROGEN CYANIDE (JACKSON, 1967)

4.8.4.1 EXTRACTION

Three gram of air-dried powdered sample was mixed thoroughly with 62.5 mL of distilled water and 3 to 4 drops of chloroform in a distillation flask. The above suspension was steam distilled. The delivery end of the condenser was kept below the surface of 5 mL of 2% KOH solution in a beaker. Approximately 30 mL of distillate was collected, and it was used as the source material.

4.8.4.2 ESTIMATION

Five mL of aliquots of the above extract were pipetted into different test tubes, and 5 mL of alkaline picrate solution was added to each test tube. The contents of the test tubes were mixed and digested in boiling water bath for 5 min, and the absorbance was measured at 520 nm against a reagent blank. The hydrogen cyanide content present in the samples was calculated by referring to a standard graph of potassium cyanide as standard and expressed as mg 100 g^{-1} of powdered samples.

4.9 STATISTICAL ANALYSIS

Proximate composition, mineral composition, vitamins (ascorbic acids and niacin), *in vitro* protein digestibility and anti-nutritional factors like total free phenolics, tannins, L-DOPA, phytic acid, hydrogen cyanide, trypsin inhibitor activities, and oligosaccharides were estimated on triplicate determinations.

CHAPTER 5

Exploration of Wild Edible Underutilized Plants

A precious gift of our nature is the wild edible plants. Most of the ethnic communities strongly depend on these plants for their day-to-day life. Wild edible plants (WEPs) refer to species that are neither cultivated nor domesticated, but are available from their wild natural habitat and used as resources of food. From the past, edible wild plants have played a very vital part in supplementing the diet of the people. The dependence on these plants has gradually declined as more exotic plants have been introduced. But many people in tribal areas still use them as a supplement for their basic need of food. Some of them are preserved for use in a dry period or sold in the rural market. But the popularity of these wild forms has recently decreased. Apart from their traditional use of food, potentially they have many advantages. They are edible and have nutritional food value, which provides minerals like sodium, potassium, magnesium, iron, calcium, phosphorus, etc. They are immune to many diseases and often used in the different formulation of 'Ayurveda' in Indian Folk medicine. They provide fibers which prevent constipation. It is believed that special attention should be paid in order to maintain and improve this important source of food supply. To achieve this, wider and sustained acceptance of wild fruits as important dietary components must be encouraged.

A scientific study of wild fruits is important for the potential sources which could be utilized at the time of shortage or during normal days or cultivated as a supply of food material for an ever increasing population. In recent decades, WEPs have been a focus of research for many ethnobotanists. Currently, there is a renewed global interest in documenting ethnobotanical information on neglected wild edible food sources (Bharucha and Pretty, 2010). In order to preserve this information for future societies, it is crucial to promote research on wild food plants, because traditional knowledge on WEPs is being eroded through acculturation and the loss of plant biodiversity along with native people and their cultural background (Asfaw, 2009). A major objective of an ethnobotanical investigation into wild food plants is the documentation of native knowledge associated with these plants.

It is to be noted that as many as 260 wild edible plants were identified as being eaten by the Palliyar/Kanikkar/Valiyan/Pulayan tribes of Southeastern Slopes of the Western Ghats, Tamil Nadu, India. They were tabulated alphabetically with Botanical name, Family name, Vernacular name (Used by the tribes) and plant parts used (Table 5.1). Wild edible plants belong to 84 families. Of which few belong to Gymnosperms and Pteridophytes and the rest belong to Angiosperms.

Based on the nature of plant parts consumed by the tribes, the plants were grouped into different categories, viz.,

- Consumption patterns of edible tubers, rhizomes, corms and root types (Table 5.2 and Plates IV to VIII)
- Consumption patterns of edible stems, piths and apical meristems (Table 5.3 and Plate X)
- Consumption pattern of edible greens (Table 5.4 and Plate IX)
- Consumption pattern of edible flowers (Table 5.5 and Plate X)
- Consumption pattern of edible unripe fruits (Table 5.6 and Plate XI)
- Consumption pattern of edible fruits (Table 5.7 and Plates XII and XIII)
- Consumption pattern of edible seeds and seed components (Table 5.8 and Plates XIV to XVI)

Of the various parts of the wild edible plants consumed by the tribes, rhizomes, corms, tubers, bulbils and root types of 44 plant species, stem, pith and apical meristem of 19 plant species, leaves (greens) of 80 plant species, flower of 14 plant species, unripe fruits/pods of 52 plant species, ripe fruits of 91 plant species and seeds and seed components of 57 plant species are consumed. The plant parts are either eaten raw or cooked and eaten as shown in Tables 5.2–5.8.

5.1 NUTRITIONAL FACTORS

5.1.1 BIOCHEMICAL ANALYSIS

The edible pith, apical meristem, tubers, rhizomes, corms, root types, leaves (greens), seeds and seed components were selected for their biochemical analyses.

TABLE 5.1 Wild Edible Plants Used By the Tribals

Sl. No.	Botanical Name	Family	Vernacular Name	Plant part used
1.	*Abelmoschus moschatus* Medik.	Malvaceae	Kattuvendai	Fleshy root, Entire unripe fruit
2.	*Acacia caesia* (L.) Willd.	Mimosaceae	Indu	Leaves
3.	*Acacia grahamii* Vajravelu	Mimosaceae	Indu	Tender leaves
4.	*Achyranthes aspera* L.	Amaranthaceae	Nayuruvi	Young shoots, Leaves
5.	*Achyranthes bidentata* Blume	Amaranthaceae	Nayuruvi	Leaves
6.	*Aegle marmelos* (L.) Correa.	Rutaceae	Vilvam	Fleshy part of the fruit
7.	*Alangium salvifolium* (Linn. F) Wangerin	Alangiaceae	Ankollam	Fruit
8.	*Allmania nodiflora* (L.) R.Br.ex Wight var. angustifolia	Amaranthaceae	Chengkumattikeerai	Leaves
9.	*Allmania nodiflora* (L.) R.Br.ex Wight var. procumbens Hook f.	Amaranthaceae	Kumattikeerai	Leaves
10.	*Alocasia macrorhiza* (L.) G. Don.	Araceae	Maraam Cheambu	Corm
11.	*Aloe vera* (L.) Burn.f.	Liliaceae	Chothukathalai	Succulent leaves
12.	*Alternanthera bettzickiana* (Regel) Nicholson	Amaranthaceae	Ponnankannikeerai	Leaves
13.	*Alternanthera sessilis* (L).R.Br.ex DC	Amaranthaceae	Ponnankannikkeerai	Leaves
14.	*Amaranthus roxburghianus* Nevski	Amaranthaceae	Araikeerai	Leaves
15.	*Amaranthus spinosus* L.	Amaranthaceae	Mullukkeerai	Tender leaves
16.	*Amaranthus tricolor* L.	Amaranthaceae	Thandankeerai	Stem, Tender leaves
17.	*Amaranthus viridis* L.	Amaranthaceae	Kuppaikeerai	Tender leaves
18.	*Amorphophallus paeoniifolius* (Dennst) *Nicolson var. campanulatus* (Blume ex Decne) Sivadasan	Araceae	Kaatuchenai	Corm
19.	*Amorphophallus sylvaticus* (Roxb.) Kunth	Araceae	Kattukarunaikeerai	Corm, Leaves

TABLE 5.1 *(Continued)*

Sl. No.	Botanical Name	Family	Vernacular Name	Plant part used
20.	*Anacardium occidentale* L.	Anacardiaceae	Kollankuttai	Fruit
21.	*Ananas comosus* (L.) Meer.	Bromeliaceae	Annashipazham	Fruit
22.	*Annona muricata* L.	Annonaceae	Mullathe	Fruit
23.	*Annona reticulata* L.	Annonaceae	Atha	Fruit
24.	*Aponogeton natans* (L.) Engler & K.	Aponogetonaceae	Kottikizhangu	Tubers
25.	*Arenga wightii* Griff.	Arecaceae	Alapanai	Pith, Apical meristem, Peduncle
26.	*Argyreia pilosa* Arn.	Convolvulaceae	Thettukkadi	Root
27.	*Artocarpus heterophyllus* Lam.	Moraceae	Pala	Unripe Perianth, Fleshy Perianth, Kernel
28.	*Artocarpus hirsutus* Linn	Moraceae	Cheenipala	Fruit
29.	*Asparagus racemosus* Willd.	Liliaceae	Neervalli	Tubers
30.	*Asystasia gangetica* (L.)T. Anderson	Acanthaceae	Mitikeerai	Leaves
31.	*Atalantia racemosa* Wight & Arn.	Rutaceae	Katuelumichai	Fleshy part of fruit, Entire unripe fruit
32.	*Atylosia scarabaeoides* (L) Benth.	Fabaceae	Kattuthuvarai	Seed
33.	*Azadirachta indica* A. Juss.	Meliaceae	Vembu	Fleshy part of the berry
34.	*Baccaurea courtallensis* Muell. Arg.	Euphorbiaceae	Moottupalam	Fruit
35.	*Bambusa arundinacea* (Retz.) Roxb.	Poaceae	Moongil	Young shoots, Seed
36.	*Basella alba* L. var. alba	Chenopodiaceae	Kattupasali	Stem, Leaves
37.	*Bauhinia purpurea* L.	Caesalpiniaceae	Mantharai	Flower buds
38.	*Begonia malabarica* Lam.	Begoniaceae	Naarayanasanjivi	Leaves
39.	*Boerhavia diffusa* L.	Nyctaginaceae	Mookaanacharana	Leaves
40.	*Boerhavia erecta* L.	Nyctaginaceae	Kuthucharana	Leaves
41.	*Bombax ceiba* L.	Bombacaceae	Ellavu	Flower

TABLE 5.1 *(Continued)*

Sl. No.	Botanical Name	Family	Vernacular Name	Plant part used
42.	*Borassus flabellifer* L.	Arecaceae	Panai	Root like fleshy cotyledons, Pith Apical meristem, Young tender leaves, Peduncle, Pulp of the tender fruit, Endosperm
43.	*Brassica juncea* (L). Czern. & Coss.	Cruciferae	Kattukadugu	Leaves
44.	*Bridelia retusa* (L.)Spreng.	Euphorbiaceae	Kadukaipalam	Fruit
45.	*Bupleurum wightii* Mukh.var. ramosissimum (Wight & Arn) Chandrabose comb.nov.	Apiaceae	Kattuseeragam	Seed
46.	*Canarium strictum* Roxb.	Burseraceae	Kungilium	Kernel
47.	*Canavalia gladiata* (Jacq.) DC.	Fabaceae	Thampattai	Tenderpod, Seed
48.	*Canavalia virosa* Wight & Arn.	Fabaceae	Kozhiavaraimotchai	Tenderpod, Seed
49.	*Canna indica* L.	Cannaceae	Kalvazhai	Rhizome
50.	*Canthium parviflorum* Lam	Rubiaceae	Periyakaarai/ Malukkaarai	Entire fruit, Tender leaves
51.	*Capparis zeylanica* L.	Capparaceae	Kaathatikaai	Entire unripe fruit, Kernel
52.	*Capsicum annuum* L.	Solanaceae	Usimilagai	Unripe fruit
53.	*Capsicum frutescens* L.	Solanaceae	Kanamillakukeerai	Leaves, Entire unripe fruit, Dry fruit
54.	*Caralluma adscendens* (Roxb.) Haw. var. attenuata (Wight) Grav. & Mayuranathan	Asclepiadaceae	Periyasirumankeerai	Stem
55.	*Caralluma lasiantha* (Wight) N.E. Br.	Asclepiadaceae	Sirumankeerai	Stem
56.	*Caralluma pauciflora* (Wight) N.E. Br	Asclepiadaceae	Kozhisilumpian	Stem
57.	*Cardiospermum microcarpa* Kunth	Sapindaceae	Periyamudakkathan	Leaves
58.	*Cardiospermum canescens* Wall.	Sapindaceae	Periyamudakkathan	Leaves
59.	*Cardiospermum halicacabum* L.	Sapindaceae	Mudakkathan	Leaves

TABLE 5.1 *(Continued)*

Sl. No.	Botanical Name	Family	Vernacular Name	Plant part used
60.	*Carica papaya* L.	Caricaceae	Pappaali	Unripe fruit, Fruit
61.	*Carissa carandas* L.	Apocynaceae	Kilakkay	Fleshy part of the unripe fruit, Fleshy part of the berry
62.	*Carmona retusa* (Vahl) Masamune	Boraginaceae	Vanaguarisi	Fruit
63.	*Caryota urens* L.	Arecaceae	Koonthalpanai	Pith, Peduncle
64.	*Cassia obtusifolia* Linn.	Caesalpiniaceae	Thakara	Leaves
65.	*Cassia tora* L.	Caesalpiniaceae	Vagai	Leaves
66.	*Celosia argentea* L.	Amaranthaceae	Mavulikkeerai	Leaves
67.	*Celtis philippensis* Blanco var.wightii (Planch.) Soep	Ulmaceae	Vellaithuvari	Seed
68.	*Centella asiatica* Urb	Umbelliferae (nom. allter. Apiaceae)	Vallaarai	Leaves
69.	*Chamaecrista absus* (L.) Irwin & Barneby.	Caesalpiniaceae	Kattukanam	Seed
70.	*Chomelia asiatica* O. Kze var. rigida Gamble.	Rubiaceae	Therani	Fleshy part of the fruit
71.	*Cissus quadrangularis* L.	Vitaceae	Perandai	Stem, Leaves
72.	*Cissus vitiginea* L.	Vitaceae	Maniperandai/Perandai	Tubers, Stem, Leaves
73.	*Citrus aurantifolia* (Christm) Swingle	Rutaceae	Elumitchai	Fruit, Unripe fruit
74.	*Citrus sinensis* (L.) Osbeck	Rutaceae	Kamala organge	Fruit
75.	*Cleome gynandra* L.	Cleomaceae	Thaivalaikeerai	Leaves
76.	*Cleome viscose* L.	Cleomaceae	Naaikkadugu	Leaves
77.	*Coccinia grandis* (L) Voigt	Cucurbitaceae	Kovai	Entire unripe fruit, Leaves, Entire fruit
78.	*Cocculus hirsutus* (L.) Diels.	Menispermaceae	Vellakattukkodi	Leaves

TABLE 5.1 *(Continued)*

Sl. No.	Botanical Name	Family	Vernacular Name	Plant part used
79.	*Colocasia esculenta* (L.) Schott	Araceae	Kattuchembu	Leaves and Petiole, Corm
80.	*Commelina benghalensis* L.	Commelinaceae	Amala	Leaves
81.	*Commiphora caudata* (Wight & Arn) Engler.	Burseraceae	Mangkiluvai/ Kadikiluvai	Fleshy part of the unripe fruit, Fleshy part of the fruit
82.	*Commiphora pubescens* (Wight & Arn). Engler.	Burseraceae	Kodikilluvai	Fleshy part of the fruit, Fleshy part of the unripe fruit
83.	*Cordia obliqua* Willd. var. obliqua.	Boraginaceae	Virusu	Fleshy part of the fruit
84.	*Cordia obliqua* Willd. var. tomentosa (Wall.) Kazmi	Boraginaceae	Kalvirusu	Fleshy part of the fruit
85.	*Coriandrum sativum* L.	Apiaceae	Kothamalli	Leaves
86.	*Costus speciosus* (Koen. Ex Retz) Sm	Costaceae	Channakoova	Rhizome
87.	*Cucumis melo* L.	Cucurbitaceae	Vellani	Fruit
88.	*Cucurbita* pepo. L.	Cucurbitaceae	Poosani	Fruit
89.	*Curculigo orchioides* Gaertn.	Hypoxidaceae	Kuluthupokie	Tubers
90.	*Curcuma neilgherrensis* Wt.	Zingiberaceae	Kattukalvazhai	Rhizome
91.	*Cyamophis tetragonoloba* Taub	Fabaceae	Seeniavarai	Unripe fruit
92.	*Cycas circinalis* L.	Cycadaceae	Palechehankizhangu	Tubers, Tender leaves, Kernel
93.	*Cyphostemma setosum* (Roxb) Alston	Vitaceae	Puliamperandai	Tubers
94.	*Decalepis hamiltonii* Wight & Arn.	Periplocaceae	Mahalikizhangu	Tubers
95.	*Digera muricata* (L.) Mart	Amaranthaceae	Kattukkeerai	Leaves
96.	*Dioscorea alata* L.	Dioscoreaceae	Kattukaychil	Tubers
97.	*Dioscorea bulbifera* L. var. vera Prain & Burkill	Dioscoreaceae	Karuvalli	Tubers
98.	*Dioscorea esculenta* (Lour.) Burkill	Dioscoreaceae	Sirukizhangu	Tubers

TABLE 5.1 *(Continued)*

Sl. No.	Botanical Name	Family	Vernacular Name	Plant part used
99.	*Dioscorea hispida* Dennst	Dioscoreaceae	Chavalkizhangu	Tubers
100.	*Dioscorea oppositifolia* L. var. dukhumensis Praine & Burkill	Dioscoreaceae	Vethalaivalli	Tubers
101.	*Dioscorea oppositifolia* L. var. oppositifolia	Dioscoreaceae	Thallavalli/Kavalavalli	Tubers
102.	*Dioscorea pentaphylla* L. var. pentaphylla	Dioscoreaceae	Mullvalli	Tubers, Bulbils
103.	*Dioscorea spicata* Roth	Dioscoreaceae	Vethalaivalli	Tubers
104.	*Dioscorea tomentosa* Koen. Ex Spreng	Dioscoreaceae	Noolvalli	Tubers
105.	*Dioscorea wallichi* Hook. F	Dioscoreaceae	Neduvai	Tubers
106.	*Diospyros ferrea* (Wild) Bakh var. neilgherrensis (Wight) Bakh	Ebenaceae	Karunthuvarai	Unripe fruit, Entire fruit
107.	*Diospyros foliosa* Wall. Ex A.DC.	Ebenaceae	Thumla	Entire fruit
108.	*Diplocyclos palmatus* (L.) Jeffrey	Cucurbitaceae	Aattupudal/ Malaipusanni	Leaves, Entire fruit
109.	*Dolichos lablab* var. Vulgaris L.	Fabaceae	Motchai	Seed
110.	*Dolichos trilobus* L.	Fabaceae	Minna	Tubers, Tenderpod, Seed
111.	*Drypetes sepiaria* (Wight &Arn). Pax & Hoffm.	Euphorbiaceae	Kalvirai	Seed
112.	*Eclipta alba* Hassk	Asteraceae	Karisalankanni	Young shoots
113.	*Elaeocarpus tectorius* (Lour.) Poir	Elaeocarpaceae	Kotla	Kernel, Fleshy part of the fruit
114.	*Eleusine coracana* (L.) Gaertn.	Poaceae	Kattukepai	Seed
115.	*Emblica officinalia* Gaertn	Euphorbiaceae	Malainelli	Fruit
116.	*Emilia sonchifolia* (L).DC	Asteraceae	Yelthanikkeerai	Leaves
117.	*Ensete superbum* (Roxb). Cheesman	Musaceae	Malaivazhai	Seed, Skinned off unripe fruit, Flower, Fleshy part of the fruit

TABLE 5.1 *(Continued)*

Sl. No.	Botanical Name	Family	Vernacular Name	Plant part used
118.	*Entada rheedi* Spreng.	Mimosaceae	Malamthellukka	Kernel
119.	*Erythroxylon monogynum* Roxb.	Erythroxylaceae	Chemmana	Fleshy part of the fruit
120.	*Eugenia caryophyllata* Wight	Myrtaceae	Kirambu	Flower buds
121.	*Euphorbia hirta* L.	Euphorbiaceae	Ammanpacharisi	Leaves
122.	*Ficus benghalensis* L. var. benghalensis.	Moraceae	Aal	Entire fruit
123.	*Ficus racemosa* L.	Moraceae	Atthi	Entire fruit
124.	*Ficus religiosa* L.	Moraceae	Arasu	Entire fruit
125.	*Flacourtia indica* (Burm.f.)Merr.	Flacourtiaceae	Mullumayilai	Fleshy part of the fruit
126.	*Flueggea leucopyrus* Willd	Euphorbiaceae	Pulanji	Fruit
127.	*Gardenia gummifera* L.f.	Rubiaceae	Karadivetchi	Fruit
128.	*Gardenia resinifera* Roth	Rubiaceae	Vetchi	Fleshy part of the fruit
129.	*Gisekia pharnaceoides* L.	Molluginaceae	Manalkeerai	Leaves
130.	*Givotia rottleriformis* Griff.	Euphorbiaceae	Vandalai	Kernel
131.	*Glinus oppositifolius* (L) A	Molluginaceae	Pachankeerai	Leaves
132.	*Glycosmis pentaphylla* (Retz.) DC.	Rutaceae	Panam Palam/ Pannichedi	Fleshy part of the fruit
133.	*Grewia flavescens* Juss.	Tiliaceae	Odaachu	Fleshy part of the fruit
134.	*Grewia heterotricha* Mast.	Tiliaceae	Periyaachu	Fleshy part of the fruit
135.	*Grewia hirsuta* Vahl	Tiliaceae	Chinnaachu	Fleshy part of the fruit
136.	*Grewia laevigata* Vahl	Tiliaceae	Karuachu	Fleshy part of the fruit
137.	*Grewia tiliifolia* Vahl	Tiliaceae	Valukkaimaram	Fleshy part of the fruit
138.	*Grewia villosa* Willd.	Tiliaceae	Vattachi	Fleshy part of the fruit
139.	*Hemidesmus indicus* (L.) R. Br. var. indicus	Periplocaceae	Nannari	Root

TABLE 5.1 *(Continued)*

Sl. No.	Botanical Name	Family	Vernacular Name	Plant part used
140.	*Hemidesmus indicus* (L.) R. Br. var. pubescens (Wight & Arn.) Hook.f.	Periplocaceae	Nannari	Root
141.	*Heracleum rigens* Wall. ex DC. var. rigens	Apiaceae	Kattukothamalli	Seed, Entire unripe fruit, Entire dry fruit, Leaves
142.	*Hibiscus lobatus* (Murr.) Kuntze	Malvaceae	Kattuvendai	Entire unripe fruit
143.	*Hibiscus lunariifolius* Willd.	Malvaceae	Vendai	Entire unripe fruit
144.	*Hibiscus ovalifolius* (Forsk.) Vahl	Malvaceae	Theingaapottu	Entire unripe fruit
145.	*Hybanthus enneaspermus* (L.)Fv.Muell.	Violaceae	Orithalthamarai	Leaves
146.	*Impatiens balsamina* L.	Balsaminaceae	Aivartenkittumpai	Seed
147.	*Ipomoea alba* L.	Convolvulaceae	Mukkuthikkay	Swollen pedicle, Young shots
148.	*Ipomoea aquatica* Forssk	Convolvulaceae	Sarkaraivalli/ Kattukothamalli	Young shoots, Leaves
149.	*Ipomoea pes-tigridis* L	Convolvulaceae	Kottalikkeerai	Leaves
150.	*Ipomoea staphylina* Roem. & Sch.	Convolvulaceae	Oanakodi	Root
151.	*Jasminum auriculatum* Vahl	Oleaceae	Kattumullai	Leaves
152.	*Jasminum calophyllum* Wall.ex.A.DC	Oleaceae	Kattumullai	Leaves
153.	*Kalanchoe pinnata* (Lam) Pers.	Crassulaceae	Rannakalli	Leaves
154.	*Kedrostis foetidissima* (Jacq.) cogn	Cucurbitaceae	Appakovai	Tubers
155.	*Kirganelia reticulata* (Poir). Baill	Euphorbiaceae	Polan	Fruit
156.	*Lablab purpureus* (L) Sweet var. lignosus (Prain) Kumari comb.	Fabaceae	Kattumotchai	Seed, Entire tender pod
157.	*Lablab purpureus* (L.) Sweet var. purpureus	Fabaceae	Avarai	Unripe pod, Seed
158.	*Lantana camara* L. var.aculeata (L)Mold.	Verbenaceae	Unni	Entire fruit
159.	*Lantana indica* Roxh	Verbenaceae	Unni	Fruit

TABLE 5.1 *(Continued)*

Sl. No.	Botanical Name	Family	Vernacular Name	Plant part used
160.	*Lathyrus sativus* L.	Fabaceae	Patani	Young shoots, Leaves
161.	*Leucas montana* Spr.var wightii HK.F	Labiatae (non alter Lamiaceae)	Aduppusattikkerai	Leaves
162.	*Luffa acutangula* (L.) Roxb. var. amara Clarke	Cucurbitaceae	Kattupirkku	Skinned off unripe fruit
163.	*Lycopersicon esculentum* Mill	Solanaceae	Chinnathakkali	Unripe fruit, Fruit
164.	*Macrotyloma uniflorum* (Lam.) Verdac	Fabaceae	Kanam	Seed
165.	*Maerua oblongifolia* (Forsk) A. Rich	Capparaceae	Boomisarkarai Kizhangu	Tubers
166.	*Mangifera indica* L.	Anacardiaceae	Maa	Fleshy part of the unripe fruit, Fleshy part of the fruit
167.	*Manihot esculenta* Crantz	Euphorbiaceae	Maracheeni	Tubers
168.	*Maranta arundinacea* L.	Marantaceae	Koovaikilangu	Rhizome
169.	*Miliusa eriocarpa* Dunn.	Annonaceae	Nedunaarai	Fleshy part of the fruit
170.	*Mimusops elengi* L.	Sapotaceae	Mahilam	Fleshy part of the fruit
171.	*Mollugo pentaphylla* Linn.	Molluginaceae	Kozhuppacheera	Leaves
172.	*Momordica charantia* L. var. charantia	Cucurbitaceae	Kuruvithalai Paakakai	Entire unripe fruit, Fleshy part of the fruit
173.	*Momordica dioica* Roxb. ex Wild	Cucurbitaceae	Poompavai/ Palupaakakai	Tubers, Entire unripe fruit, Fleshy part of the fruit
174.	*Moringa concanensis* Nimmo ex Gibs.	Moringaceae	Kattumoringai	Flower, Entire unripe fruit, Kernel, Leaves
175.	*Mucuna atropurpurea* DC.	Fabaceae	Thellukka	Kernel
176.	*Mucuna pruriens* (L.) DC. var. pruriens	Fabaceae	Poonaikali	Seed

TABLE 5.1 *(Continued)*

Sl. No.	Botanical Name	Family	Vernacular Name	Plant part used
177.	*Mucuna pruriens* (L.) DC. var. utilis (Wall ex wight) Baker ex. Burck. (Black coloured seed coat)	Fabaceae	Poonaikali	Seed
178.	*Mucuna pruriens* (L.) DC. var. utilis (Wall ex wight) Baker ex. Burck. (White coloured seed coat)	Fabaceae	Poonaikali	Seed
179.	*Mukia maderaspatana* (L.) M	Cucurbitaceae	Musumusukkai	Tender Leaves, Entire fruit
180.	*Murraya koenigii* (L.) Spreng.	Rutaceae	Kariveppilai	Leaves, Flower, Fleshy part of the fruit
181.	*Murraya paniculata* (L). Jack	Rutaceae	Kattukariveppilai	Leaves, Flower, Fleshy part of the fruit
182.	*Neonotonia wightii* (Wight & Arn) Lackey. var. coimbatorensis (Ajita sen) Karthik.	Fabaceae	Kattumotchai	Seed
183.	*Nephrolepis auriculata* (L.) Trimen	Polypodiaceae	Neerveelchi	Tubers
184.	*Nymphaea pubescens* Willd	Nymphacaceae	Allikizhangu	Tubers
185.	*Nymphaea rubra* Roxb. ex Andrews	Nymphaeaceae	Tamaray Kizhangu	Tubers
186.	*Ocimum gratissimum* L.	Labiatae (non alter. Lamiaceae)	Kattuthulasi	Seed
187.	*Opuntia dillenii* (Ker-Gawl.) Haw	Cactaceae	Sappathikkalli	Fleshy part of the fruit
188.	*Oryza meyeriana* (Zoll. & Mor.) Baill var. granulata (Nees & Arn.ex Watt) Duist.	Poaceae	Nell	Grain (Rice)
189.	*Osyris quadripartita* Salzm.ex Decne. var. puberula (Hook.f.) Kumari.	Santalaceae	Sundaravalli	Entire fruit
190.	*Oxalis corniculata* L.	Oxalidaceae	Puliyotharakeerai	Leaves
191.	*Oxalis latifolia* Kunth	Oxalidaceae	Pulichangkeerai	Leaves

TABLE 5.1 *(Continued)*

Sl. No.	Botanical Name	Family	Vernacular Name	Plant part used
192.	*Parthenocissus neilgherriensis* (Wight) Planch	Vitaceae	Kattukarunai Kizhangu	Tubers
193.	Passiflora foetida L.	Passifloraceae	Poonakkaali	Entire fruit
194.	*Pavetta erassicaulis* Bremek	Rubiaceae	Pavattanchedi	Tender unripe fruit
195.	*Pavetta indica* L. var. indica	Rubiaceae	Pavattai	Tender unripe fruit
196.	*Peperomia pellucida* (L) H.B.K	Piperaceae	Thippili	Leaves
197.	*Phoenix pusilla* Gaertn.	Arecaceae	Iechamaram	Apical meristem, Tender inflorescence, Fleshy part of the unripe fruit (red in colour), Fleshy part of the fruit
198.	*Phyllanthus emblica* L.	Euphorbiaceae	Kattunelli	Fleshy part of the unripe fruit, Fleshy part of the fruit
199.	*Phyllanthus reticulatus* Poir.	Euphorbiaceae	Poola	Entire fruit
200.	*Physalis minima* L. var. indica Clarke	Solanaceae	Kuttythakkali	Leaves, Fleshy part of the Fruit
201.	*Piper longum* L.	Piperaceae	Thippili	Spike
202.	*Pithecellobium dulce* (Roxb.) Beneth.	Mimosaceae	Kodukkapuli	Aril
203.	*Polyalthia cerasoides* (Roxb.) Bedd.	Annonaceae	Nedunarai	Fleshy part of the fruit
204.	*Polyalthia suberosa* (Roxb.) Thw.	Annonaceae	Kodinaaval	Fleshy part of the fruit
205.	*Portulaca oleracea* L. var. oleracea	Portulacaceae	Udumbukoluppukeerai	Leaves
206.	*Portulaca quadrifida* L.	Portulacaceae	Paruppukeerai	Leaves
207.	*Premna corymbosa* (Burn.f.)Rottl. & Willd.	Verbenaceae	Minnakkeerai	Leaves
208.	*Psidium guajava* L.	Myrtaceae	Kattukoyya	Entire unripe fruit, Fleshy part of the fruit
209.	*Psilanthus wightianus* (Wight & Arn.) J.	Rubiaceae	Kattumalli	Tender leaves
210.	*Rhynchosia cana* DC.	Fabaceae	Kattuthuvarai	Seed

TABLE 5.1 *(Continued)*

Sl. No.	Botanical Name	Family	Vernacular Name	Plant part used
211.	*Rhynchosia filipes* Benth.	Fabaceae	Kattuthuvarai	Seed
212.	*Rhynchosia rufescens* (Willd.) DC.	Fabaceae	Kattuthuvarai	Seed
213.	*Rhynchosia suaveolens* (L.f) DC	Fabaceae	Kattuthuvarai	Seed
214.	*Rubus ellipticus* Sm	Rosaceae	Karunganni	Fruit
215.	*Rubus niveus* Thumb.var. niveus Gamble	Rosaceae	Maekattuillanthai	Entire fruit
216.	*Rubus racemosus* Roxh	Rosaceae	Cheetipalam	Fruit
217.	*Sapindus emarginatus* Vahl	Sapindaceae	Pullichi	Seed
218.	*Sarcostemma acidum* (Roxb) Voigt	Asclepiadaceae	Thannikkodi/ Kodikkalli	Leaves, Stem, Tubers
219.	*Scutia myrtina* (Burm.f) Kurz	Rhamnaceae	Thorattipalam	Fruit
220.	*Secamone emetica* (Retz.) R.Br.ex Schultes	Asclepiadaceae	Karuppattikodi	Entire unripe fruit, Fleshy part of the fruit
221.	*Senna occidentalis* (L.) Link.	Caesalpiniaceae	Ponnavarai/Tagarai	Tenderpod
222.	*Sesamum indicum* L.	Pedaliaceae	Kattu yellu	Seed
223.	*Sesbania grandiflora* (L.) Poir.	Fabaceae	Agathikeerai	Leaves
224.	*Shorea roxburghii* GDon	Dipterocarpaceae	Kungilium	Cotyledons
225.	*Solanum anguivi* Lam. var. multiflora (Roth ex. Roem & Sch.) Chithra comb. nov.	Solanaceae	Kattuthudhuvalai	Leaves, Entire unripe fruit
226.	*Solanum erianthum* D.Don	Solanaceae	Kattuchundai	Entire unripe fruit
227.	*Solanum melongena* L. var. insanum (L) Prain.	Solanaceae	Mullukathari	Spine removed unripe fruit
228.	*Solanum nigrum* L.	Solanaceae	Milaguthakkali	Leaves, Unripe fruit, Entire fruit
229.	*Solanum pubescens* Willd.	Solanaceae	Kattuchundai	Entire unripe fruit
230.	*Solanum torvum* Sw.	Solanaceae	Kattuchundai	Entire unripe fruit

TABLE 5.1 *(Continued)*

Sl. No.	Botanical Name	Family	Vernacular Name	Plant part used
231.	*Solanum trilobatum* L	Solanaceae	Thudhuvalai	Leaves, Entire unripe fruit, Entire fruit
232.	*Sterculia foetida* L.	Sterculiaceae	Kongatti	Kernel
233.	*Sterculia guttata* Roxb. ex DC.	Sterculiaceae	Kattuiluppai	Kernel
234.	*Sterculia urens* Roxb.	Sterculiaceae	Vennaali	Root, Exudates from the stem, Kernel
235.	*Streblus asper* Lour.	Moraceae	Pira	Latex from the stem
236.	*Strychnos nux-vomica* L.	Loganiaceae	Theathankottai	Kernel
237.	*Syzygium cumini* (L.) Skeels.	Myrtaceae	Naval	Fleshy part of the fruit
238.	*Syzygium jambos* (L.) Alston	Myrtaceae	Perunaaval	Fruit
239.	*Tamarindus indica* L.	Caesalpiniaceae	Puli	Leaves, Flower, Fleshy part of the unripe fruit, Fleshy part of the fruit, Kernel
240.	*Teramnus labialis* (L.f) Spreng.	Fabaceae	Kattukanam	Seed
241.	*Terminalia bellirica* (Gaertn.) Roxb.	Combretaceae	Tani	Kernel
242.	*Terminalia chebula* Retz.	Combretaceae	Kadukkai	Kernel, Fruit
243.	*Tinospora cordifolia* (Willd) Miers ex HookF.& Thomas	Menispermaceae	Cheenthil	Leaves
244.	*Trianthema portulacastrum* L.	Aizoaceae	Vattachanathikeerai	Leaves
245.	*Trichosanthes tricuspidata*. Lour var. tricuspidata	Cucurbitaceae	Korattaipalam	Fruit
246.	*Uvaria rufa* Blume.	Annonaceae	Thevakodi	Fleshy part of the fruit
247.	*Vaccinium leschenaultii* Wt. var. rotundifolia Cl.	Vacciniaceae	Kalavu	Fruit
248.	*Vigna bourneae* Gamble	Fabaceae	Kattu payaru	Seed/Entire pod/Unripe Seed
249.	*Vigna radiata* (L.) Wilczek var. sublobata (Roxb.) Verdc.	Fabaceae	Kattupasipayaru	Seed, Entire tender pod

TABLE 5.1 *(Continued)*

Sl. No.	Botanical Name	Family	Vernacular Name	Plant part used
250.	*Vigna trilobata* (L.) Verdc.	Fabaceae	Kalpayaru	Entire tender pod, Seed
251.	*Vigna unguiculata* (L.) Walp.subsp. cylindrica (L) Eselt.	Fabaceae	Panni Minnppayaru	Unripe pod, Seed
252.	*Vigna unguiculata* (L.) Walp.subsp. unguiculata.	Fabaceae	Kattukanam	Seed
253.	*Xanthosoma sagittifolium* (L.) Schott	Araceae	Paalecheambu	Corm
254.	*Xanthosoma violaceum* Schott	Araceae	Neelacheambu	Corm
255.	*Xylia xylocarpa* (Roxb.) Taub.	Mimosaceae	Iral	Kernel
256.	*Ziziphus mauritiana* Lam var. mauritiana	Rhamnaceae	Ilanthai	Fruit
257.	*Ziziphus mauritiana* Lam. var. fruticosa (Haines) Seb & Balakr.	Rhamnaceae	Periyailanthai	Fleshy part of the fruit, Fleshy part of the unripe fruit
258.	*Ziziphus oenopola* (L) Mill.	Rhamnaceae	Pulichi	Fleshy part of the fruit
259.	*Ziziphus rugosa* Lam	Rhamnaceae	Mullukottai	Fruit, Kernel, Fleshy part of the unripe fruit
260.	*Ziziphus xylopyrus* (Retz.) Willd.	Rhamnaceae	Mullukottai	Fleshy part of the fruit

TABLE 5.2 Consumption Pattern of Edible Tubers, Rhizomes, Corms and Root Types By the Tribals

Sl. No.	Botanical Name	Family	Vernacular Name	Ethnic Community	Plant part used	Mode of use
1.	*Abelmoschus moschatus* Medik.	Malvaceae	Kattuvendai	Palliyar Pulayan	Fleshy root	Raw/Boiled/Roasted
2.	*Amorphophallus paeoniifolius* (Dennst) Nicolson var. *campanulatus* (Blume ex Decne) Sivadasan	Araceae	Kaatuchenai	Kanikkar	Corm	Boiled
3.	*Amorphophallus sylvaticus* (Roxb.) Kunth	Araceae	Kattukarunai Keerai	Palliyar Pulayan Kanikkar	Corm	Boiled in Tamarindus water and consumed next day
4	*Aponogeton natans* (L.) Engler & K.	Aponogetonaceae	Kottikizhangu	Valaiyan	Tubers	Raw/Boiled
5.	*Alocasia macrorhiza* (L.) G. Don.	Araceae	Maraam Cheambu	Kanikkar	Corm	Boiled
6.	*Argyreia pilosa* Arn.	Convolvulaceae	Thettukkadi	Palliyar Pulayan	Root	Boiled/Roasted
7.	*Asparagus racemosus* Willd.	Liliaceae	Neervalli	Palliyar Pulayan Kanikkar Valaiyan	Tubers	Raw/Boiled/Roasted
8.	*Borassus flabellifer* L.	Arecaceae	Panai	Palliyar Pulayan Kanikkar	Root like fleshy cotyledons	Boiled/Roasted
9.	*Canna indica* L.	Cannaceae	Kalvazhai	Palliyar Pulayan Valaiyan	Rhizome	Boiled/Roasted

TABLE 5.2 *(Continued)*

Sl. No.	Botanical Name	Family	Vernacular Name	Ethnic Community	Plant part used	Mode of use
10	*Cissus vitiginea* L.	Vitaceae	Maniperandai	Valaiyan	Tubers	Boiled
11.	*Colocasia esculenta* (L.) Schott.	Araceae	Kattuchembu	Palliyar Pulayan Kanikkar Valaiyan	Corm	Boiled/ Roasted/ Cooked with chilli and salt as curry
12.	*Costus speciosus* (Koen. Ex Retz) Sm	Costaceae	Channakoova	Palliyar	Rhizome	Cooked with salt, chilli, tamarind and turmeric powder and used as curry
13.	*Curculigo orchioides* Gaertn.	Hypoxidaceae	Kuluthupokie	Kanikkar Valaiyan	Tubers	Raw/Boiled/Roasted
14.	Curcuma neilgherrensis Wt.	Zingiberaceae	Kattukalvazhai	Palliyar Pulayan	Rhizome	Raw/Boiled/Roasted
15.	*Cycas circinalis* L.	Cycadaceae	Palechehan Kizhangu	Valaiyan	Tubers	Boiled
16	*Cyphostemma setosum* (Roxb) Alston	Vitaceae	Puliamperandai	Valaiyan	Tubers	Boiled
17.	*Decalepis hamiltonii* Wight & Arn.	Periplocaceae	Mahalikizhangu	Valaiyan	Tubers	Cut into pieces add chilli and salt to make pickles
18.	*Dioscorea alata* L.	Dioscoreaceae	Kattukaychil	Kanikkar	Tubers	Boiled/Roasted
19.	*Dioscorea bulbifera* L. var. *vera* Prain & Burkill	Dioscoreaceae	Karuvalli	Palliyar Pulayan	Tubers	Boiled/Roasted
20.	*Dioscorea esculenta* (Lour.) Burkill	Dioscoreaceae	Sirukizhangu	Kanikkar	Tubers	Boiled/Roasted
21.	*Dioscorea hispida* Dennst	Dioscoreaceae	Chavalkizhangu	Palliyar	Tubers	Boiled/Roasted

TABLE 5.2 *(Continued)*

Sl. No.	Botanical Name	Family	Vernacular Name	Ethnic Community	Plant part used	Mode of use
22.	*Dioscorea oppositifolia* L. var. *dukhumensis* Praine Burkill	Dioscoreaceae	Vethalaivalli	Palliyar Pulayan	Tubers	Boiled/Roasted
23	*Dioscorea oppositifolia* L. var. *oppositifolia*	Dioscoreaceae	Thallavalli/ Kavalavalli	Palliyar Pulayan Valaiyan	Tubers	Boiled/Roasted
24.	*Dioscorea pentaphylla* L. var *pentaphylla*	Dioscoreaceae	Mullvalli	Palliyar Pulayan Valaiyan	Tubers Bulbils	Boiled/Roasted
25.	*Dioscorea tomentosa* Koen. Ex Spreng	Dioscoreaceae	Noolvalli	Palliyar Pulayan Kanikkar	Tubers	Tender part of the tubers boiled/Roasted and eaten by Palliyar tribals in Ayyanarkoil
26	*Dioscorea spicata* Roth	Dioscoreaceae	Vethalaivalli	Valaiyan	Tubers	Boiled/Roasted
27.	*Dioscorea wallichii* Hook. F	Dioscoreaceae	Neduvai	Kanikkar	Tubers	Boiled/Roasted
28.	*Dolichos trilobus* L.	Fabaceae	Minna	Palliyar	Tubers	Boiled/Roasted
29.	*Hemidesmus indicus* (L.) R. Br. var. *Indicus*	Periplocaceae	Nannari	Palliyar Pulayan Kanikkar Valaiyan	Root	Juice extracted, mixed with water and sugar
30.	*Hemidesmus indicus* (L.) R. Br. var. *pubescens* (Wight & Arn.) Hook.f.	Periplocaceae	Nannari	Palliyar Pulayan	Root	Juice extracted, mixed with water and sugar
31.	*Ipomoea staphylina* Roem. & Sch.	Convolvulaceae	Oanakodi	Palliyar Pulayan Valaiyan	Root	Raw/Boiled

TABLE 5.2 *(Continued)*

Sl. No.	Botanical Name	Family	Vernacular Name	Ethnic Community	Plant part used	Mode of use
32	*Kedrostis foetidissima* (Jacq.) cogn	Cucurbitaceae	Appakovai	Valaiyan	Tubers	Boiled/Roasted
33.	*Maerua oblongifolia* (Forsk) A. Rich	Capparaceae	Boomisarkarai Kizhangu	Valaiyan	Tubers	Boiled
34.	*Manihot esculenta* Crantz	Euphorbiaceae	Maracheeni	Kanikkar	Tubers	Boiled/Roasted
35	*Maranta arundinacea* L.	Marantaceae	Koovaikilangu	Kanikkar	Rhizome	Boiled/Roasted
36.	*Momordica dioica* Roxb. ex Wild	Cucurbitaceae	Poompavai	Valaiyan	Tubers	Boiled/Roasted
37.	*Nephrolepis auriculata* (L.) Trimen	Polypodiaceae	Neerveelchi	Palliyar Pulayan	Tubers	Boiled
38.	*Nymphaea pubescens* Willd	Nymphacaceae	Allikizhangu	Valaiyan Kanikkar	Tuber	Boiled
39.	*Nymphaea rubra* Roxb. ex Andrews	Nymphaeaceae	Tamaray Kizhangu	Valaiyan Kanikkar	Tubers	Boiled
40.	*Parthenocissus neilgherriensis* (Wight) Planch	Vitaceae	Kattukarunai Kizhangu	Valaiyan Kanikkar	Tubers	Boiled in tamarindus water and consumed
41.	*Sarcostemma acidum* Roxb. Voigt	Asclepiadaceae	Kodikkalli	Valaiyan Kanikkar	Tubers	Raw
42.	*Sterculia urens* Roxb.	Sterculiaceae	Vennaali	Palliyar Pulayan	Root	Raw/Boiled
43.	*Xanthosoma sagittifolium* (L.) Schott	Araceae	Paalecheambu	Kanikkar	Corm	Boiled/Roasted
44.	*Xanthosoma violaceum* Schott	Araceae	Neelacheambu	Kanikkar	Corm	Boiled/Roasted

TABLE 5.3 Consumption Pattern of Edible Stems, Piths, and Apical Meristems By the Tribals

Sl. No.	Botanical Name	Family	Vernacular Name	Ethnic Community	Plant part used	Mode of use
1.	*Achyranthes aspera* L.	Amaranthaceae	Nayuruvi	Kanikkar	Young shoots	Boiled young shoots are eaten
2.	*Amaranthus tricolor* L.	Amaranthaceae	Thandankeerai	Palliyar, Pulayan, Valaiyan	Stem	Boiled with chilli and salt
3.	*Arenga wightii* Griff.	Arecaceae	Alapanai	Palliyar Pulayan	Pith	Powdered pith used in the preparation of 'dosai'/bread. Powdered pith mixed with sugar used in preparation of sweets
					Apical meristem	Raw/Cut into pieces and boiled
4.	*Bambusa arundinacea* (Retz.) Roxb.	Poaceae	Moongil	Kanikkar	Young shoots	Young shoots are eaten as such
5.	*Basella alba* L. var. *alba*	Chenopodiaceae	Kattupasali	Palliyar Pulayan	Stem	Boiled along with salt and chilli
6.	*Borassus flabellifer* L.	Arecaceae	Panai	Palliyar Pulayan	Pith Apical meristem	Raw
7.	*Caralluma adscendens* (Roxb.) Haw. var. *attenuata* (Wight) Grav. & Mayuranathan	Asclepiadaceae	Periyasirumankeerai	Palliyar Pulayan Valaiyan	Stem	Raw
8.	*Caralluma lasiantha* (Wight) N.E. Br.	Asclepiadaceae	Sirumankeerai	Palliyar Pulayan	Stem	Raw
9.	*Caralluma pauciflora* (Wight) N.E. Br	Asclepiadaceae	Kozhisilumpian	Valaiyan	Stem	Raw

TABLE 5.3 *(Continued)*

Sl. No.	Botanical Name	Family	Vernacular Name	Ethnic Community	Plant part used	Mode of use
10.	*Caryota urens* L.	Arecaceae	Koonthalpanai	Palliyar Pulayan	Pith	Raw
11.	*Cissus quadrangularis* L.	Vitaceae	Perandai	Palliyar Pulayan Kanikkar	Stem	Peeled stem mashed and used in preparation of chutney and side dishes
12.	*Cissus vitiginea* L.	Vitaceae	Perandai	Palliyar Pulayan	Stem	Peeled stem is used to make chutney/cooked as a food item
13.	*Eclipta alba* Hassk	Asteraceae	Karisalan Kanni	Kanikkar	Young shoots	Boiled young shoots are eaten
14.	*Ipomoea aquatica* Forssk	Convolvulaceae	Sarkaraivalli	Kanikkar	Young shoots	Boiled young shoots are eaten
15.	*Lathyrus sativus* L.	Fabaceae	Patani	Kanikkar	Young shoots	Boiled young shoots are eaten
16.	*Phoenix pusilla* Gaertn.	Arecaceae	Iechamaram	Palliyar Pulayan Kanikkar	Apical meristem	Raw
17.	*Sarcostemma acidum* (Roxb.) Voigt	Asclepiadaceae	Thannikkodi	Palliyar	Stem	Raw
18.	*Sterculia urens* Roxb.	Sterculiaceae	Vennaali	Palliyar Pulayan	Exudates from the stem	Mixed with water and salt – kept over night and used.
19.	*Streblus asper* Lour.	Moraceae	Pira	Palliyar	Latex from the stem	Mixed with milk and sugar used in preparation of sweet

TABLE 5.4 Consumption Pattern of Edible Greens By the Tribals

Sl. No.	Botanical Name	Family	Vernacular Name	Ethnic Community	Plant part used	Mode of use
1.	*Acacia caesia* (L.) Willd.	Mimosaceae	Indu	Palliyar Pulayan	Leaves	Boiled and mashed with salt and chilli
2.	*Acacia grahamii* Vajravelu	Mimosaceae	Indu	Palliyar Pulayan	Tender leaves	Boiled and mashed with salt and chilli
3.	*Achyranthes aspera* L.	Amaranthaceae	Nayuruvi	Kanikkar	Leaves	Boiled young leaves are eaten
4.	*Achyranthes bidentata* Blume	Amaranthaceae	Nayuruvi	Palliyar Pulayan	leaves	Boiled and mashed with salt and chilli
5.	*Allmania nodiflora* (L.) R.Br.ex Wight var *angustifolia*	Amaranthaeae	Chengkumattikeerai	Palliyar	Leaves	Boiled and mashed with salt and chilli
6.	*Allmania nodiflora* (L.) R.Br.ex Wight var *procumbens* Hook f.	Amaranthaeae	Kumattikeerai	Palliyar Valaiyan	Leaves	Boiled and mashed with salt and chilli
7.	*Aloe vera* (L.) Burn.f.	Liliaceae	Chothukathalai	Palliyar Pulayan	Succulent leaves	Peeled and cooked with dhal, tomato and salt as curry
8.	*Alternanthera bettzickiana* (Regel) Nicholson	Amaranthaceae	Ponnankannikeerai	Palliyar Pulayan	Leaves	Fried along with chilli coconut and salt.
9.	*Alternanthera sessilis* (L).R.Br.ex DC	Amaranthaceae	Ponnankannikkeerai	Palliyar Pulayan Valaiyan	Leaves	Fried along with chilli, coconut and salt.
10.	*Amaranthus roxburghianus* Nevski	Amaranthaceae	Araikeerai	Palliyar Pulayan	Leaves	Boiled and mashed with salt and chilli
11.	*Amaranthus spinosus* L.	Amaranthaceae	Mullukkeerai	Palliyar Pulayan	Tender leaves	Boiled and mashed with salt and chilli

TABLE 5.4 *(Continued)*

Sl. No.	Botanical Name	Family	Vernacular Name	Ethnic Community	Plant part used	Mode of use
12.	*Amaranthus tricolor* L.	Amaranthaceae	Thandankeerai	Palliyar Pulayan Valaiyan	Tender leaves	Boiled and mashed with salt and chilli
13.	*Amaranthus viridis* L.	Amaranthaceae	Kuppaikeerai	Palliyar Pulayan Valaiyan	Tender leaves	Boiled and mashed with salt and chilli
14.	*Amorphophallus sylvaticus* (Roxb.) Kunth	Araceae	Kattukarunaikeerai	Palliyar Pulayan	Leaves	Boiled and mashed along with more tamarindus water
15.	*Asystasia gangetica* (L.)T. Anderson	Acanthaceae	Mitikeerai	Palliyar, Kanikar	Leaves	Boiled and mashed with salt and chilli
16.	*Basella alba* L.var. *alba*	Chenopodiaceae	Kattupasali	Palliyar Pulayan	Leaves	Boiled and mashed with salt and chilli
17.	*Begonia malabarica* Lam.	Begoniacceae	Naarayanasanjivi	Palliyar Kanikkar	Leaves	Raw
18.	*Boerhavia diffusa* L.	Nyctaginaceae	Mookaanacharana	Palliyar Pulayan	Leaves	Boiled and mashed with salt and chilli
19.	*Boerhavia erecta* L.	Nyctaginaceae	Kuthucharana	Palliyar Pulayan Valaiyan	Leaves	Boiled and mashed with salt and chilli
20.	*Borassus flabellifer* L.	Arecaceae	Panai	Palliyar Pulayan	Young tender leaves	Raw
21.	*Brassica juncea* (L). Czern.& Coss.	Cruciferae	Kattukadugu	Palliyar Pulayan Valaiyan	Leaves	Boiled and mashed with salt and chilli

TABLE 5.4 *(Continued)*

Sl. No.	Botanical Name	Family	Vernacular Name	Ethnic Community	Plant part used	Mode of use
22.	*Canthium parviflorum* Lam.	Rubiaceae	Periyakaarai/ Malukkaarai	Palliyar	Tender leaves	Raw /Boiled and mashed with salt and chilli
23.	*Cassia obtusifolia* Linn.	Caesalpiniaceae	Thakara	Kanikkar	Leaves	Boiled and mashed with salt and chilli
24.	*Cassia tora* L.	Caesalpiniaceae	Vagai	Kanikkar Valaiyan	Leaves	Boiled young leaves are eaten
25.	*Capsicum frutescens* L.	Solanaceae	Kanamillakukeerai	Palliyar Pulayan	Leaves	Boiled and mashed with salt and chilli
26	*Capsicum annuum*	Solanaceae	Millakukeerai	Palliyar Kanikar	Leaves	Boiled and mashed with salt and chilli
27.	*Cardiospermum canescens* Wall.	Sapindaceae	Periyamudakkathan	Palliyar Pulayan	Leaves	Boiled and mashed with salt and chilli
28.	*Cardiospermum halicacabum* L.	Sapindaceae	Mudakkathan	Palliyar Pulayan Valaiyan	Leaves	Boiled and mashed with salt and chilli
29.	*Cardiospermum microcarpa* Kunth	Sapindaceae	Periyamudakkathan	Valaiyan	Leaves	Boiled and mashed with salt and chilli
30.	*Celosia argentea* L.	Amaranthaceae	Mavulikkeerai	Palliyar Pulayan Valaiyan	Leaves	Boiled and mashed with salt and chilli
31.	*Centella asiatica* Urb	Umbelliferae (nom.alltes. Apiaceae)	Vallaarai	Palliyar Pulayan Valaiyan	Leaves	Boiled and mashed with salt and chilli
32.	*Cissus quadrangularis* L.	Vitaceae	Perandai	Palliyar Pulayan	Leaves	Boiled and mashed with salt and chilli

TABLE 5.4 *(Continued)*

Sl. No.	Botanical Name	Family	Vernacular Name	Ethnic Community	Plant part used	Mode of use
33	*Cissus vitiginea* L.	Vitaceae	Perandai	Palliyar Pulayan	Leaves	Boiled and mashed with salt and chilli
34.	*Cleome gynandra* L.	Cleomaceae	Thaivalaikeerai	Palliyar Pulayan Kanikkar Valaiyan	Leaves	Boiled and mashed with salt and chilli
35.	*Cleome viscosa* L.	Cleomaceae	Naaikkadugu	Palliyar Pulayan Kanikkar Valaiyan	Leaves	Boiled and mashed with salt and chilli
36.	*Coccinia grandis* (L.) Voigt	Cucurbitaceae	Kovai	Palliyar Pulayan Valaiyan	Leaves	Boiled and mashed with salt and chilli
37.	*Cocculus hirsutus* (L.) Diels.	Menispermaceae	Vellakattukkodi	Palliyar Valaiyan	Leaves	Extract poured into sugar solution and the semisolid from consumed
38.	*Colocasia esculenta* (L.) Schott	Araceae	Kattuchembu	Palliyar Pulayan Valaiyan	Leaves and Petiole	Boiled and mashed with chilli, salt and coconut
39.	*Commelina benghalensis* L.	Commelinaceae	Amala	Palliyar Pulayan	Leaves	Boiled and mashed with chilli and salt
40.	*Commelina ensifolia* R.Br.	Commelinaceae	Amala	Palliyar Pulayan	Leaves	Boiled and mashed with chilli and salt
41.	*Coriandrum sativum* L.	Apiaceae	Kothamalli	Kanikkar	Leaves	Leaves are uses as condiment.

TABLE 5.4 *(Continued)*

Sl. No.	Botanical Name	Family	Vernacular Name	Ethnic Community	Plant part used	Mode of use
42.	*Cycas circinalis* L.	Cycadaceae	Paereenji	Palliyar	Tender leaves	Decanted leaves mashed with chilli and salt
43.	*Digera muricata* (L.) Mart	Amaranthaceae	Kattukkeerai	Palliyar Pulayan Valaiyan	Leaves	Boiled and mashed with chilli and salt
44.	*Diplocyclos palmatus* (L.) Jeffrey	Cucurbitaceae	Aattupudal/ Malaipusanni	Valaiyan	Leaves	Boiled and mashed with chilli and salt
45	*Emilia sonchifolia* (L).DC	Asteraceae	Yelthanikkeerai	Kanikkar	Leaves	Boiled Leaves are eaten
46.	*Euphorbia hirta* L.	Euphorbiaceae	Ammanpachcharisi	Kanikkar	Leaves	Boiled Leaves are eaten
47.	*Gisekia pharnaceoides* L.	Molluginaceae	Manalkeerai	Palliyar Pulayan	Leaves	Boiled and mashed with chilli and salt
48.	*Glinus oppositifolius* (L)A	Molluginaceae	Pachankeerai	Valaiyan	Leaves	Boiled and mashed with chilli and salt
49.	*Heracleum rigens* Wall ex Dc var. rigens	Apiaceae	Kattukothamalli	Palliyar Pulayan	Leaves	Fried and mashed along with chilli, tamarindus and salt
50.	*Hybanthus enneaspermus* (L.)Fv.Muell.	Violaceae	Orithalthamarai	Palliyar Valaiyan	Leaves	Raw
51.	*Ipomoea aquatica* Lorssk	convolvalaceae	Kattukothamalli	Kanikkar	Leaves	Young Leaves are eaten as such
52.	*Ipomoea pes-tigridis* L	Convolvulaceae	Kottalikkeerai	Valaiyan	Leaves	-
53.	*Jasminum auriculatum* Vahl	Oleaceae	Kattumullai	Palliyar	Leaves	Boiled and mashed with chilli and salt
54.	*Jasminum calophyllum* Wall.ex.A.DC	Oleaceae	Kattumullai	Palliyar	Leaves	Boiled and mashed with chilli and salt

TABLE 5.4 *(Continued)*

Sl. No.	Botanical Name	Family	Vernacular Name	Ethnic Community	Plant part used	Mode of use
55.	*Kalanchoe pinnata* (Lam) Pers.	Crassulaceae	Rannakalli	Palliyar Pulayan	Leaves	Raw/Boiled/Fried along with salt chilli and coconut.
56.	*Lathyrus sativus* L.	Fabaceae	Patani	Kanikkar	Leaves	Boiled Leaves are eaten
57.	*Leucas montana* Spr.var *wightii* HK.F	Labiatae (non alter Lamiaceae)	Aduppusattikkerai	Palliyar Pulayan	Leaves	Boiled/Fried with coconut chilli and salt
58.	*Mollugo pentaphylla* Linn.	Molluginaceae	Kozhuppacheera	Palliyar Kanikkar	Leaves	Boiled and mashed with salt and chilli
59.	*Moringa concanensis* Nimmo ex Gibs.	Moringaceae	Kattumoringai	Palliyar Pulayan Valaiyan	Leaves	Boiled/Fried along with salt chilli and coconut.
60.	*Mukia maderaspatana* (L.) M	Cucurbitaceae	Musumusukkai	Palliyar Pulayan Valaiyan	Tender leaves	Raw
61.	*Murraya koenigii* (L.) Spreng.	Rutaceae	Kariveppilai	Palliyar Pulayan	Leaves	Mashed with chilli salt and piece of coconut (Chutney)
62.	*Murraya paniculata* (L). Jack	Rutaceae	Kattukariveppilai	Palliyar Pulayan	Leaves	Mashed along with chilli salt and piece of coconut (Chutney)
63.	*Oxalis corniculata* L.	Oxalidaceae	Puliyotharakeerai	Palliyar/ Pulayan/Val	Leaves	Boiled and mashed along with chilli and salt
64.	*Oxalis latifolia* Kunth	Oxalidaceae	Pulichangkeerai	Palliyar Pulayan	Leaves	Boiled and mashed along with chilli and salt
65.	*Peperomia pellucida* (L) H.B.K	Piperaceae	Thippili	Kanikkar	Leaves	Leaves are eaten as such

TABLE 5.4 *(Continued)*

Sl. No.	Botanical Name	Family	Vernacular Name	Ethnic Community	Plant part used	Mode of use
66.	*Physalis minima* L. var. *indica* Clarke	Solanaceae	Kuttythakkali	Palliyar Pulayan Valaiyan	Leaves	Boiled and mashed along with chilli and salt
67.	*Portulaca oleracea* L. var. *oleracea*	Portulacaceae	Udumbukoluppukeerai	Palliyar	Leaves	Boiled and mashed along with chilli and salt
68.	*Portulaca quadrifida* L.	Portulacaceae	Paruppukeerai	Palliyar Pulayan	Leaves	Boiled and mashed along with salt and chilli (taken only by ladies)
69.	*Premna corymbosa* (Burn.f.)Rottl. & Willd.	Verbenaceae	Minnakkeerai	Palliyar Pulayan	Leaves	Boiled and mashed along with chilli and salt
70.	*Psilanthus wightianus* (Wight & Arn.) J.	Rubiaceae	Kattumalli	Palliyar	Tender leaves	Fried along with chilli and salt
71.	*Sarcostemma acidum* (Roxb) Voigt	Asclepiadaceae	Thannikkodi	Palliyar	Leaves	Raw
72.	*Sesbania grandiflora* (L.) Poir.	Fabaceae	Agathikeerai	Palliyar Pulayan	Leaves	Fried along with chilli coconut and salt
73.	*Solanum anguivi* Lam. var. *multiflora* (Roth ex. Roem & Sch.) Chithra comb.nov.	Solanaceae	Kattuthudhuvalai	Palliyar Pulayan	Leaves	Boiled and mashed with salt and chilli
74.	*Solanum nigrum* L.	Solanaceae	Milaguthakkali	Palliyar Pulayan Valaiyan	Leaves	Boiled and mashed with salt and chilli
75.	*Solanum trilobatum* L	Solanaceae	Thudhuvalai	Palliyar	Leaves	Boiled and mashed with salt and chilli

TABLE 5.4 *(Continued)*

Sl. No.	Botanical Name	Family	Vernacular Name	Ethnic Community	Plant part used	Mode of use
76.	*Tamarindus indica* L.	Caesalpiniaceae	Puli	Palliyar Pulayan Valaiyan	Leaves	Raw
77.	*Tinospora cordifolia* (Willd) Miers ex HookF.& Thomas	Menispermaceae	Cheenthil	Kakkan	Leaves	Boiled leaves are eaten
78.	*Trianthema portulacastrum* L.	Aizoaceae	Vattachanathikeerai	Palliyar Pulayan Valaiyan	Leaves	Fried along with chilli, coconut and salt.
79.	*Vigna radiata* var. *sublobata* (Roxb.) Verdc.	Fabaceae	Kattuparipayaru	Palliyar	Leaves	Boiled and mashed with salt and chilli
80.	*Vigna trilobata* (L.)Verdc.	Fabaceae	Kalpayaru	Palliyar	Leaves	Boiled and mashed with salt and chilli

TABLE 5.5 Consumption Pattern of Edible Flowers By the Tribals

Sl. No.	Botanical Name	Family	Vernacular Name	Ethnic Community	Plant part used	Mode of use
1.	*Arenga wightii* Griff.	Arecaceae	Alapanai	Palliyar	Peduncle	Toddy tapped from the peduncle
2.	*Bauhinia purpurea* L.	Caesalpiniaceae	Mantharai	Kanikkar	Flower buds	Flower buds are cooked and eaten
3.	*Bombax ceiba* L.	Bombacaceae	Ellavu	Valaiyan Kanikkar	Flower	Flower buds are used as vegetable
4.	*Borassus flabellifer* L.	Arecaceae	Panai	Palliyar	Peduncle	Toddy tapped from the peduncle
5.	*Caryota urens* L.	Arecaceae	Koonthalpanai	Palliyar Kanikkar	Peduncle	Toddy tapped from the peduncle
6.	*Ensete superbum* (Roxb.) Cheesman.	Musaceae	Malaivazhai	Palliyar	Flower	Cut into pieces and boiled with chilli and coconut
7.	*Eugenia caryophyllata* Wight	Myrtaceae	Kirambu	Valaiyan Kanikkar	Flower buds	Flower buds are used as spices
8.	*Ipomoea alba* L.	Convolvulaceae	Mukkuthikkay	Palliyar	Swollen pedicle	Fried
9.	*Moringa concanensis* Nimmo ex Gibs.	Moringaceae	Kattumoringai	Palliyar Valaiyan	Flower	Fried in oil along with coconut
10.	*Murraya koenigii* (L.) Spreng.	Rutaceae	Kariveppilai	Palliyar Kanikkar	Flower	Fried along with salt, chilli and coconut oil
11.	*Murraya paniculata* (L.) Jack	Rutaceae	Kattu kariveppilai	Palliyar Kanikkar	Flower	Fried along with salt, chilli and coconut oil
12.	*Phoenix pusilla* Gaertn.	Areaceae	Iechamaram	Palliyar	Tender inflorescence	Fried along with salt, chilli and coconut oil
13.	*Piper longum* L.	Piperaceae	Thippili	Kanikkar Valaiyan	Spike	Spike eaten as masticatory
14.	*Tamarindus indica* L.	Caesalpiniaceae	Puli	Palliyar Kanikkar Valaiyan	Flower	Raw/Mashed along with chilli and salt

TABLE 5.6 Consumption Pattern of Edible Unripe Fruits By the Tribals

Sl. No.	Botanical Name	Family	Vernacular Name	Ethnic Community	Plant part used	Mode of use
1.	*Abelmoschus moschatus* Medik.	Malvaceae	Kattuvendai	Palliyar Pulayan	Entire unripe fruit	Raw/Fried in oil along with chilli and salt
2.	*Artocarpus heterophyllus* Lam.	Moraceae	Pala	Palliyar Pulayan	Unripe Perianth	Cut into pieces and boiled with chilli and coconut
3.	*Atalantia racemosa* Wight & Arn.	Rutaceae	Katuelumichai	Palliyar Pulayan	Entire unripe fruit	Pickled in salt and chilli powder/Juice extracted, mixed with water and a sugar
4.	*Canavalia gladiata* (Jacq.)DC.	Fabaceae	Thampattai	Palliyar Pulayan	Tenderpod	Boiled with chilli and salt
5.	*Canavalia virosa* Wight & Arn.	Fabaceae	Kozhiavaraimotchai	Palliyar	Tenderpod	Tender pods are boiled with chilli and salt as a food item
6.	*Capparis zeylanica* L.	Capparaceae	Kaathatikaai	Palliyar Pulayan	Entire unripe fruit	Raw/Boiled/Fried
7.	*Capsicum frutescens* L.	Solanaceae	Kanamillaku keerai	Palliyar Pulayan	Entire unripe fruit	Raw
8.	*Capsicum annuum* L.	Solanaceae	Usimilagai	Valaiyan	Unripe fruit	Cut into pieces mashed with salt and coconut to make chutney.
9.	*Carissa carandas* L.	Apocynaceae	Kilakkay	Palliyar Pulayan	Fleshy part of the unripe fruit	Pickled in salt and chilli powder
10.	*Carica papaya* L.	Caricaceae	Pappaali	Palliyar	Unripe fruit	Fleshy part of the Unripe fruit cut into the pieces are boile and with coconut, chilli and salt as a food item.
11.	*Citrus aurantifolia* (Christm)Swingle.	Rutaceae	Elumitchai	Palliyar	Unripe fruit	Chopped unripe fruits are make into pickle with chilli powder and salt.
12.	*Coccinia grandis* (L) Voigt	Cucurbitaceae	Kovai	Palliyar Pulayan	Entire unripe fruit	Unripe fruits sun dried and fried.

TABLE 5.6 *(Continued)*

Sl. No.	Botanical Name	Family	Vernacular Name	Ethnic Community	Plant part used	Mode of use
13.	*Commiphora caudata* (Wight & Arn) Engler.	Burseraceae	Mangkiluvai	Palliyar Pulayan	Fleshy part of the unripe fruit	Raw
14.	*Commiphora pubescens* (Wight & Arn) Engler.	Burseraceae	Kadikiluvai	Palliyar	Fleshy part of the unripe fruit	Raw
15.	*Cyamophis tetragonoloba* Taub	Fabaceae	Seeniavarai	Palliyar Pulayan	Unripe fruit	Unripe fruits cut into pieces and fried with chilli and salt as a food item
16.	*Diospyros ferrea* (Wild) Bakh var. *neilgherrensis* (Wight) Bakh	Ebenaceae	Karunthuvarai	Palliyar Pulayan	Unripe fruit	Unripe fruits are taken as food into natural state cooked with chilli and salt as a food item
17.	*Dolichos trilobus* L.	Fabaceae	Minnaa	Palliyar	Tenderpod	Raw/Boiled
18.	*Ensete superbum* (Roxb.) Cheesman	Musaceae	Malaivazhai	Palliyar Pulayan	Skinned off unripe fruit	Boiled unripe fruit fried along with coconut, salt and chilli
19.	*Heracleum rigens* Wall. Ex DC. var. *rigens*	Apiaceae	Kattukotamalli	Palliyar Pulayan	Entire unripe fruit	Boiled and mashed along withi chilli and salt
20.	*Hibiscus lobatus* (Murr.) Kuntze	Malvaceae	Kattuvendai	Palliyar Pulayan	Entire unripe fruit	Raw
21.	*Hibiscus lunariifolius* Willd.	Malvaceae	Vendai	Palliyar Pulayan	Entire unripe fruit	Raw

TABLE 5.6 *(Continued)*

Sl. No.	Botanical Name	Family	Vernacular Name	Ethnic Community	Plant part used	Mode of use
22.	*Hibiscus ovalifolius* (Forsk.) Vahl	Malvaceae	Theingaapottu	Palliyar	Entire unripe fruit	Raw
23.	*Lablab purpureus* (L.) Sweet var. *lignosus* (Prain) Kumari comb.	Fabaceae	Kattumochai	Palliyar Pulayan	Entire tender pod	Boiled
24.	*Lablab purpureus* (L.) Sweet var. *purpureus*	Fabaceae	Avarai	Valaiyan	Unripe pod	Boiled
25.	*Lycopersicon esculentum* Mill	Solanaceae	Chinnathakkali	Palliyar Pulayan	Unripe fruit	Unripe fruits are taken as food in their natural state cooked with chilli and salt as a food item.
26.	*Luffa acutangula* (L.) Roxb. var. *amara* Clarke	Cucurbitaceae	Kattu pirkku	Palliyar Pulayan	Skinned off unripe fruit	Boiled along with chilli and salt.
27.	*Mangifera indica* L.	Anacardiaceae	Maa	Palliyar Pulayan Kanikkar	Fleshy part of the unripe fruit	Raw
28.	*Momordica charantia* L. var. *charantia*	Cucurbitaceae	Kuruvithalai Paakakai	Palliyar Pulayan Kanikkar	Entire unripe fruit	Cut and oil-fried with chilli, onion and salt
29.	*Momordica dioica* Roxb.ex Willd.	Cucurbitaceae	Palupaakakai	Palliyar Pulayan	Entire unripe fruit	Cut and oil-fried with chilli, onion and salt

TABLE 5.6 *(Continued)*

Sl. No.	Botanical Name	Family	Vernacular Name	Ethnic Community	Plant part used	Mode of use
30.	*Moringa concanensis* Nimmo ex Gibs.	Moringaceae	Kattumoringai	Palliyar Pulayan Kanikkar	Entire unripe fruit	Fried in oil with salt and coconut
31.	*Pavetta erassicaulis* Bremek	Rubiaceae	Pavattanchedi	Valaiyan	Tender unripe fruit	Boiled
32.	*Pavetta indica* L. var. indica	Rubiaceae	Pavattai	Palliyar Pulayan	Tender unripe fruit	Fried in oil with salt and chilli
33.	*Phoenix pusilla* Gaertn.	Arecaceae	Iechamaram	Palliyar Pulayan Kanikkar	Fleshy part of the unripe fruit (red in colour)	Raw
34.	*Phyllanthus emblica* L.	Euphorbiaceae	Kattunelli	Palliyar Pulayan Kanikkar Valaiyan	Fleshy part of the unripe fruit	Raw
35.	*Psidium guajava* L.	Myrtaceae	Kattu koyya	Palliyar Pulayan Kanikkar Valaiyan	Entire unripe fruit	Raw
36.	*Secamone emetica* (Retz.) R.Br.ex Schulted	Asclepiadaceae	Karuppattikodi	Palliyar Pulayan	Entire unripe fruit	Raw
37.	*Senna occidentalis* (L.) Link.	Caesalpiniaceae	Ponnavarai/Tagarai	Palliyar Pulayan	Tenderpod	Fried along with salt, coconut and chilli

TABLE 5.6 *(Continued)*

Sl. No.	Botanical Name	Family	Vernacular Name	Ethnic Community	Plant part used	Mode of use
38.	*Solanum anguivi* Lam. var. *multiflora* (Roth ex Roem.& Sch.) Chithra comb.nov.	Solanaceae	Kattu thudhuvalai	Palliyar Pulayan	Entire unripe fruit	Fried
39.	*Solanum erianthum* D.Don	Solanaceae	Kattu chundai	Palliyar Pulayan	Entire unripe fruit	Fried in oil
40	*Solanum melongena* L. var. *insanum* (L) Prain.	Solanaceae	Mullukathari	Palliyar Pulayan	Spine removed unripe fruit	Fried along with dry fish salt and chilli
41	*Solanum nigrum* L.	Solanaceae	Milaguthakkali	Palliyar Pulayan Kanikkar Valaiyan	Unripe fruit	Unripe fruit are taken as food in their natural state cooked with chilli and salt as a food item.
42.	*Solanum pubescens* Willd.	Solanaceae	Kattuchundai	Palliyar Pulayan	Entire unripe fruit	Sun dried/Roasted used in curry
43.	*Solanum torvum* Sw.	Solanaceae	Kattuchundai	Palliyar Pulayan Kanikkar	Entire unripe fruit	Sun dried/Roasted used in curry
44.	*Solanum trilobatum* L.	Solanaceae	Thudhuvalai	Palliyar Pulayan Kanikkar Valaiyan	Entire unripe fruit	Fried in oil

TABLE 5.6 *(Continued)*

Sl. No.	Botanical Name	Family	Vernacular Name	Ethnic Community	Plant part used	Mode of use
45.	*Tamarindus indica* L.	Caesalpiniaceae	Puli	Palliyar Pulayan	Fleshy part of the unripe fruit	Raw
46	*Vigna bourneae* Gamble.	Fabaceae	Kattu payaru	Palliyar Pulayan	Entire pod/ Unripe seed	Raw /Boiled
47.	*Vigna radiata* (L.) Wilczek var. *sublobata*	Fabaceae	Kattu pasipayaru	Palliyar Pulayan	Entire tender pod	Raw /Boiled
48.	*Vigna trilobata* (L.) Verdc.	Fabaceae	Kalpayaru	Palliyar Pulayan	Entire tender pod	Raw/Boiled
49.	*Vigna unguiculata* (L.) Walp.subsp. *cylindrica* (L). Eselt.	Fabaceae	Panni Minnappayaru	Palliyar Pulayan	Unripe pod	Raw/Boiled
50.	*Ziziphus mauritiana* Lam. var. *fruticosa* (Haines) Seb.&Balakr.	Rhamnaceae	Periya ilanthai	Palliyar Pulayan	Fleshy part of the unripe fruit	Raw
51.	*Ziziphus rugosa* Lam.	Rhamnaceae	Mullukottai	Palliyar Pulayan	Fleshly part of the unripe fruit	Raw
52.	*Ziziphus xylopyrus* (Retz.) Willd.	Rhamnaceae	Mullukottai	Palliyar Pulayan	Fleshy part of the unripe fruit	Raw

TABLE 5.7 Consumption Pattern of Edible Fruits By the Tribals

Sl. No.	Botanical Name	Family	Vernacular Name	Ethnic Community	Plant part used	Mode of use
1.	*Aegle marmelos* (L.) Correa.	Rutaceae	Vilvam	Palliyar Pulayan	Fleshy part of the fruit	Raw
2.	*Alangium salvifolium* (Linn. F) Wangerin	Alangiaceae	Ankollam	Palliyar	Fruit	Raw
3.	*Anacardium occidentale* L.	Anacardiaceae	Kollankuttai	Kanikkar	Fruit	The raw fruits are salted and eaten
4.	*Ananas comosus* (L.) Meer.	Bromeliaceae	Annashipazham	Palliyar Pulayan	Fruit	Fleshy part of the fruit is taken as such as food
5.	*Annona muricata* L.	Annonaceae	Mullathe	Kanikkar	Fruit	Raw
6.	*Annona reticulata* L.	Annonacae	Atha	Kanikkar Palliyar	Fruit	Raw
7.	*Artocarpus heterophyllus* Lam.	Moraceae	Pala	Palliyar Pulayan	Fleshy perianth	Raw
8.	*Artocarpus hirsutus* Linn	Moraceae	Cheenipala	Palliyar	Fruit	Raw
9.	*Atalantia racemosa* Wight & Arn.	Rutaceae	Kattuelumichai	Palliyar Pulayan	Fleshy part of the fruit	Raw
10.	*Azadirachta indica* A. Juss.	Meliaceae	Vembu	Palliyar Pulayan	Fleshy part of the berry	Raw
11.	*Baccaurea courtallensis* Muell. Arg.	Euphorbiaceae	Moottupalam	Kanikkar	Fruit	Raw fruits are prepare as pickle
12.	*Borassus flabellifer* L.	Arecaceae	Panai	Palliyar Pulayan	Pulp of the tender fruit	Raw
13.	*Bridelia retusa* (L.) Spreng.	Euphorbiaceae	Kadukaipalam	Palliyar Pulayan	Fruit	Raw

TABLE 5.7 *(Continued)*

Sl. No.	Botanical Name	Family	Vernacular Name	Ethnic Community	Plant part used	Mode of use
14.	*Canthium parviflorum* Lam	Rubiaceae	Periyakaarai/ Malukkaarai	Palliyar	Entire fruit	Raw
15.	*Capsicum frutescens* L.	Solanaceae	Kanamillaku keerai	Palliyar	Dry fruit	Powdered and used in curry preparation.
16.	*Carmona retusa* (Vahl) Masamune	Boraginaceae	Vanaguarisi	Palliyar	Fruit	Raw
17.	*Carica papaya* L.	Caricaceae	Pappaali	Palliyar Pulayan	Fruit	Raw
18.	*Carissa carandas* L.	Apocynaceae	Kilakkay	Palliyar	Fleshy part of the berry	Raw
19.	*Chomelia asiatica* O. Kze var. *rigida* Gamble.	Rubiaceae	Therani	Palliyar	Fleshy part of the fruit	Raw
20.	*Citrus aurantifolia* (Christm) Swingle	Rutaceae	Elumitchai	Palliyar Pulayan	Fruit	Fruit juice mixed with water and sugar is taken as a soft drink to regrets.
21.	*Citrus sinensis* (L.) Osbeck	Rutaceae	Kamala organge	Palliyar Pulayan	Fruit	Juicy leaves of the fruit are taken as food.
22.	*Coccinia grandis* (L.) Voigt	Cucurbitaceae	Kovai	Palliyar	Entire fruit	Raw
23.	*Commiphora caudata* (Wight & Arn). Engler	Burseraceae	Mangkilluvai	Palliyar	Fleshy part of the fruit	Raw
24.	*Commiphora pubescens* (Wight & Arn). Engler.	Burseraceae	Kodikilluvai	Palliyar	Fleshy part of the fruit	Raw
25.	*Cordia obliqua* Willd. var. *obliqua.*	Boraginaceae	Virusu	Palliyar Pulayan	Fleshy part of the fruit	Raw

TABLE 5.7 *(Continued)*

Sl. No.	Botanical Name	Family	Vernacular Name	Ethnic Community	Plant part used	Mode of use
26.	*Cordia obliqua* Willd. var. *tomentosa* (Wall.) Kazmi	Boraginaceae	Kalvirusu	Palliyar	Fleshy part of the fruit	Raw
27.	*Cucumis melo* L.	Cucurbitaceae	Vellani	Kanikkar	Fruit	Raw fruits are eaten
28.	*Cucurbita pepol.* L.	Cucurbitaceae	Poosani	Kanikkar	Fruit	Cooled fruits are eaten
29.	*Diospyros ferrea* (Willd.) Bakh. Var. *neilgherrensis* (Wight) Bakh.	Ebenaceae	Karunthuvarai	Palliyar	Entire fruit	Raw
30.	*Diospyros foliosa* Wall. Ex A.DC.	Ebenaceae	Thumla	Palliyar	Entire fruit	Raw
31.	*Diplocyclos palmatus* (L.) Jeffrey	Cucurbitaceae	Aattupudal/ Malaipusanni	Palliyar Pulayan	Entire fruit	Raw
32.	*Elaeocarpus tectorius* (Lour.) Poir.	Elaeocarpaceae	Kotla	Palliyar	Fleshy part of the fruit	Raw
33.	*Emblica officinalia* Gaertn	Euphorbiaceae	Malainelli	Palliyar Kanikkar	Fruit	Raw
34.	*Ensete superbum* (Roxb.) Cheesman.	Musaceae	Malaivazhai	Palliyar Pulayan	Fleshy part of the fruit	Raw
35.	*Erythroxylon monogynum* Roxb.	Erythroxylaceae	Chemmana	Palliyar	Fleshy part of the fruit	Raw
36.	*Ficus benghalensis* L. var. *benghalensis*.	Moraceae	Aal	Palliyar Pulayan	Entire fruit	Raw
37.	*Ficus racemosa* L.	Moraceae	Atthi	Palliyar Pulayan	Entire fruit	Raw/Dipped in honey and dried
38.	*Ficus religiosa* L.	Moraceae	Arasu	Palliyar Pulayan	Entire fruit	Raw

TABLE 5.7 *(Continued)*

Sl. No.	Botanical Name	Family	Vernacular Name	Ethnic Community	Plant part used	Mode of use
39.	*Flacourtia indica* (Burm.f.)Merr.	Flacourtiaceae	Mullumayilai	Palliyar Pulayan	Fleshy part of the fruit	Raw
40.	*Flueggea leucopyrus* Willd	Euphorbiaceae		Kanikkar	Fruit	Raw
41.	*Gardenia gummifera* L.f.	Rubiaceae	Karadivetchi	Palliyar Pulayan	Fruit	Raw
42.	*Gardenia resinifera* Roth	Rubiaceae	Vetchi	Palliyar Pulayan	Fleshy part of the fruit	Raw
43.	*Glycosmis pentaphylla* (Retz.) DC.	Rutaceae	Panam Palam/ Pannichedi	Palliyar Pulayan	Fleshy part of the fruit	Raw
44.	*Grewia flavescens* Juss.	Tiliaceae	Odaachu	Palliyar	Fleshy part of the fruit	Raw
45.	*Grewia heterotricha* Mast.	Tiliaceae	Periyaachu	Palliyar	Fleshy part of the fruit	Raw
46.	*Grewia hirsuta* Vahl	Tiliaceae	Chinnaachu	Palliyar	Fleshy part of the fruit	Raw
47.	*Grewia laevigata* Vahl	Tiliaceae	Karuachu	Palliyar	Fleshy part of the fruit	Raw
48.	*Grewia tiliifolia* Vahl	Tiliaceae	Valukkaimaram	Palliyar	Fleshy part of the fruit	Raw
49.	*Grewia villosa* Willd.	Tiliaceae	Vattachi	Palliyar	Fleshy part of the fruit	Raw
50.	*Heracleum rigens* Wall.ex DC. var. *rigens.*	Apiaceae	Kattukothamalli	Palliyar Pulayan	Entire dry fruit	Roasted fruit fried mashed along with coconut, salt and chilli
51.	*Kirganelia reticulata* (Poir). Baill	Euphorbiaceae	Polan	Kanikkar	Fruit	Raw

TABLE 5.7 *(Continued)*

Sl. No.	Botanical Name	Family	Vernacular Name	Ethnic Community	Plant part used	Mode of use
52.	*Lantana camara* L. var.*aculeata* (L) Mold.	Verbenaceae	Uni	Palliyar Pulayan	Entire fruit	Raw
53.	*Lantana indica* Roxh	Verbenaceae	Unni	Palliyar Pulayan	Fruit	Raw
54.	*Lycopersicon esculentum* Mill	Solanaceae	Chinnathakkali	Palliyar Pulayan	Fruit	Raw
55.	*Mangifera indica* L.	Anacardiaceae	Maa	Palliyar Pulayan	Fleshy part of the fruit	Raw
56.	*Miliusa eriocarpa* Dunn.	Annonaceae	Nedunaarai	Palliyar	Fleshy part of the fruit	Raw
57.	*Mimusops elengi* L.	Sapotaceae	Mahilam	Palliyar Pulayan	Fleshy part of the fruit	Raw
58.	*Momordica charantia* L.var. *charantia*	Cucurbitaceae	Kuruvithalai Paakakai	Palliyar	Fleshy part of the fruit	Raw
59.	*Momordica dioica* Roxb.ex Willd.	Cucurbitaceae	Palupaakakai	Palliyar	Fleshy part of the fruit	Raw
60.	*Mukia maderaspatana* (L.) M.	Cucurbitaceae	Musumusukkai	Palliyar Pulayan	Entire fruit	Raw
61.	*Murraya koenigii* (L.)Spreng.	Rutaceae	Kariveppilai	Palliyar Pulayan	Fleshy part of the fruit	Raw
62..	*Murraya paniculata* (L.) Jack.	Rutaceae	Kattukariveppilai	Palliyar Pulayan	Fleshy part of the fruit	Raw
63.	*Opuntia dillenii* (Ker-Gawl.) Haw	Cactaceae	Sappathikkalli	Palliyar Pulayan	Fleshy part of the fruit	Raw
64.	*Osyris quadripartita* Salzm.ex Decne. var.*puberula* (Hook.f.) Kumari.	Santalaceae	Sundaravalli	Palliyar	Entire fruit	Raw

TABLE 5.7 *(Continued)*

Sl. No.	Botanical Name	Family	Vernacular Name	Ethnic Community	Plant part used	Mode of use
65.	*Passiflora foetida* L.	Passifloraceae	Poonakkaali	Palliyar	Entire fruit	Raw
66.	*Phoenix pusilla* Gaertn.	Arecaceae	Iechamaram	Palliyar Pulayan	Fleshy part of the fruit	Raw
67.	*Phyllanthus emblica* L.	Euphorbiaceae	Kattunelli	Palliyar Pulayan	Fleshy part of the fruit	Raw
68.	*Phyllanthus reticulatus* Poir.	Euphorbiaceae	Poola	Palliyar	Entire fruit	Raw
69.	*Physalis minima* L. var. *indica* Clarke.	Solanaceae	Kutty thakkali	Palliyar	Fleshy part of the Fruit	Raw
70.	*Polyalthia cerasoides* (Roxb.) Bedd.	Annonaceae	Nedunarai	Palliyar	Fleshy part of the fruit	Raw
71.	*Polyalthia suberosa* (Roxb.) Thw.	Annonaceae	Kodinaaval	Palliyar Pulayan	Fleshy part of the fruit	Raw
72.	*Psidium guajava* L.	Myrtaceae	Kattu koyya	Palliyar Pulayan	Fleshy part of the fruit	Raw
73.	*Rubus ellipticus* Sm	Rosaceae	Karunganni	Palliyar Pulayan	Fruit	Raw
74.	*Rubus niveus* Thumb.var.*niveus* Gamble	Rosaceae	Maekattu illanthai	Palliyar Pulayan	Entire fruit	Raw
75.	*Rubus racemosus* Roxh	Rosaceae	Cheetipalam	Palliyar Pulayan	Fruit	Raw
76.	*Scutia myrtina* (Burm.f) Kurz	Rhamnaceae	Thorattipalam	Valaiyan	Fruit	
77.	*Secamone emetica* (Retz.) R.Br.ex Schultes	Asclepiadaceae	Karuppattikodi	Palliyar	Fleshy part of the fruit	Raw
78.	*Solanum nigrum* L.	Solanaceae	Milaguthakkali	Palliyar Pulayan	Entire fruit	Raw

TABLE 5.7 *(Continued)*

Sl. No.	Botanical Name	Family	Vernacular Name	Ethnic Community	Plant part used	Mode of use
79.	*Solanum trilobatum* L.	Solanaceae	Thudhuvalai	Palliyar Pulayan	Entire fruit	Raw
80.	*Syzygium cumini* (L.) Skeels.	Myrtaceae	Naval	Palliyar Pulayan	Fleshy part of the fruit	Raw
81.	*Tamarindus indica* L.	Caesalpiniaceae	Puli	Palliyar Pulayan	Fleshy part of the fruit	Raw
82.	*Syzygium jambos* (L.) Alston	Myrtaceae	Perunaaval	Palliyar Pulayan	Fruit	Raw
83.	*Terminalia chebula* Retz.	Combretaceae	Kadukkai	Kanikkar	Fruit	Raw fruits are eaten
84.	*Trichosanthes tricuspidata*. Lour var. *tricuspidata*	Cucurbitaceae	Korattaipalam	Palliyar Pulayan	Fruit	Raw
85.	*Uvaria rufa* Blume.	Annonaceae	Thevakodi	Palliyar	Fleshy part of the fruit	Raw
86.	*Vaccinium leschenaultii* Wt. var. *rotundifolia* Cl.	Vacciniaceae	Kalavu	Palliyar Pulayan	Fruit	Raw
87.	*Ziziphus mauritiana* Lam. var. *fruticosa* (Haines) Seb & Balakr.	Rhamnaceae	Periyailanthai	Palliyar Pulayan	Fleshy part of the fruit	Raw
88.	*Ziziphus mauritiana* Lam var. *mauritiana*	Rhamnaceae	Ilanthai	Palliyar Pulayan	Fruit	Raw
89.	*Ziziphus oenopola* (L) Mill.	Rhamnaceae	Pulichi	Palliyar	Fleshy part of the fruit	Raw
90.	*Ziziphus rugosa* Lam	Rhamnaceae	Mullukottai	Palliyar Pulayan	Fruit	Raw
91.	*Ziziphus xylopyrus* (Retz.) Willd.	Rhamnaceae	Mullukottai	Palliyar Pulayan	Fleshy part of the fruit	Raw

TABLE 5.8 Consumption Pattern of Edible Seeds and Seed Components By the Tribals

Sl. No.	Botanical Name	Family	Vernacular Name	Ethnic Community	Plant part used	Mode of use
1.	*Artocarpus heterophyllus* Lam.	Moraceae	Pala	Palliyar Pulayan Kanikkar Valaiyan	Kernel	Boiled/Roasted
2.	*Atylosia scarabaeoides* (L) Benth.	Fabaceae	Kattuthuvarai	Palliyar Pulayan	Seed	Raw/Boiled/Roasted
3.	*Bambusa arundinacea* (Retz.) Roxb.	Poaceae	Moongil	Kanikkar	Seed	Seeds made into edible flour and cakes
4.	*Borassus flabellifer* L.	Arecaceae	Panai	Palliyar Pulayan Kanikkar	Endosperm	Raw
5.	*Bupleurum wightii* Mukh.var. *ramosissimum* (Wight & Arn) Chandrabose comb.nov.	Apiaceae	Kattuseeragam	Palliyar	Seed	Side dish is prepared along with salt, chilli and coconut
6.	*Canarium strictum* Roxb.	Burseraceae	Kungilium	Palliyar Pulayan Kanikkar	Kernel	Raw/Roasted
7.	*Canavalia gladiata* (Jacq.)DC.	Fabaceae	Thampattai	Palliyar Pulayan Kanikkar Valaiyan	Seed	Seeds consumed after repeated boiling
8.	*Canavalia virosa* Wight& Arn	Fabaceae	Kozhiavaraimotchai	Palliyar Pulayan	Seed	Seeds are taken as food after boiling
9.	*Capparis zeylanica* L.	Capparaceae	Kaathatikaai	Palliyar Pulayan Kanikkar Valaiyan	Kernel	Raw/Roasted
10.	*Celtis philippensis* Blanco var. *wightii* (Planch.) Soep	Ulmaceae	Vellaithuvari	Palliyar Pulayan	Seed	Raw/Roasted

TABLE 5.8 *(Continued)*

Sl. No.	Botanical Name	Family	Vernacular Name	Ethnic Community	Plant part used	Mode of use
11.	*Chamaecrista absus* (L.) Irwin & Barneby.	Caesalpiniaceae	Kattukanam	Palliyar Pulayan	Seed	Boiled/Roasted
12.	*Cycas circinalis* L.	Cycadaceae	Paereenji	Palliyar Kanikkar	Kernel	Powdered decoated dry seeds used in the preparation of various items like 'dosai'/'puttu' etc
13.	*Dolichos lablab* var. *vulgaris* L.	Fabaceae	Motchai	Palliayr	Seed	Seeds are taken as food after boiling
14.	*Dolichos trilobus* L.	Fabaceae	Minna	Palliyar Pulayan Kanikkar Valaiyan	Seed	Raw/Boiled/Roasted
15.	*Drypetes sepiaria* (Wight &Arn). Pax & Hoffm.	Euphorbiaceae	Kalvirai	Palliyar	Seed	Raw
16.	*Elaeocarpus tectorius* (Lour.) Poir	Elaeocarpaceae	Kotla	Palliyar Kanikkar	Kernel	Raw/Roasted
17.	*Eleusine coracana* (L.) Gaertn.	Poaceae	Kattu kepai	Palliyar Pulayan	Seed	Raw/Boiled
18.	*Ensete superbum* (Roxb). Cheesman	Musaceae	Malai vazhai	Palliyar Pulayan Kanikkar Valaiyan	Seed	Roasted
19.	*Entada rheedi* Spreng.	Mimosaceae	Malam thellukka	Palliyar Kanikkar	Kernel	Kernels roasted and soaked for 3 days.

TABLE 5.8 *(Continued)*

Sl. No.	Botanical Name	Family	Vernacular Name	Ethnic Community	Plant part used	Mode of use
20.	*Givotia rottleriformis* Griff.	Euphorbiaceae	Vandalai	Palliyar Pulayan Kanikkar	Kernel	Milky juice extracted from the powder of raw kernels
21.	*Heracleum rigens* Wall. ex DC. Var.*rigens*	Apiaceae	Kattukothamalli	Palliyar	Seed	Fried seeds are ground along with coconut, chilli and kari leaf
22.	*Impatiens balsamina* L.	Balsaminaceae	-	Kanikkar	Seed	Raw seeds are edible
23.	*Lablab purpureus* (L) Sweet var. *lignosus* (Prain) Kumari comb.	Fabaceae	Kattumotchai	Palliyar Pulayan Kanikkar	Seed	Roasted/Boiled
24.	*Lablab purpureus* (L.) Sweet var. *purpureus*	Fabaceae	Avari seed	Valaiyan	Seed	Seeds are taken as food after boiling
25.	*Macrotyloma unifllorum* (Lam.) Verdac	Fabaceae	Kanam	Palliyar Kanniar	Seed	Seeds are taken as food after boiling
26.	*Moringa concanensis* Nimmo ex Gibs.	Moringaceae	kattumoringai	Palliyar Pulayan Kanikkar Valaiyan	Kernel	Raw kernels cooked with chilli and salt as curry
27.	*Mucuna atropurpurea* DC.	Fabaceae	Thellukka	Palliyar Pulayan	Kernel	Roasted
28.	*Mucuna pruriens* (L.) DC. var. *pruriens*	Fabaceae		Palliyar	Seed	Seeds are soaked and boiled/Roasted and eaten as such
29.	*Mucuna pruriens* (L.) DC. var. *utilis* (Wall ex wight) Baker ex. Burck. (Black coloured seed coat)	Fabaceae	Poonaikali	Kanikkar	Seed	Seeds are soaked and boiled/Roasted and eaten as such

TABLE 5.8 *(Continued)*

Sl. No.	Botanical Name	Family	Vernacular Name	Ethnic Community	Plant part used	Mode of use
30.	*Mucuna pruriens* (L.) DC. var. *utilis* (Wall ex wight) Baker ex. Burck. (White coloured seed coat)	Fabaceae	Poonaikali	Kanikkar	Seed	Seeds are soaked and boiled/Roasted and eaten as such
31.	*Neonotonia wightii* (Wight & Arn) Lackey. var. *coimbatorensis* (Ajita sen) Karthik.	Fabaceae	Kattumotchai	Palliyar Pulayan Kanikkar Valaiyan	Seed	Raw/Boiled/Roasted
32.	*Ocimum gratissimum* L.	Labiatae	Kattuthulasi	Palliyar	Seed	Boiled
33.	*Oryza meyeriana* (Zoll. & Mor.) Baill var. *granulata* (Nees & Arn. ex Watt) Duist.	Poaceae	Nell	Palliyar	Grain (Rice)	Raw/Boiled with salt
34.	*Pithecellobium dulce* (Roxb.) Beneth.	Mimosaceae	Kodukkapuli	Palliyar Pulayan Kanikkar Valaiyan	Aril	Raw
35.	*Rhynchosia cana* DC.	Fabaceae	Kattuthuvarai	Palliyar Pulayan	Seed	Raw/Boiled/Roasted
36.	*Rhynchosia filipes* Benth.	Fabaceae	Kattuthuvarai	Palliyar	Seed	Raw/Boiled/Roasted
37.	*Rhynchosia rufescens* (Willd.) DC.	Fabaceae	Kattuthuvarai	Palliyar Pulayan	Seed	Raw/Boiled/Roasted
38.	*Rhynchosia suaveolens* (L.f) DC	Fabaceae	Kattuthuvarai	Palliyar Pulayan	Seed	Raw/Boiled/Roasted
39.	*Sapindus emarginatus* Vahl	Sapindaceae	Pullichi	Palliyar	Seed	Raw mostly children prefer
40.	*Sesamum indicum* L.	Pedaliaceae	Kattu yellu	Palliyar Pulayan Kanikkar Valaiyan	Seed	Raw/Roasted ground along with salt and red chilli/with Jaggary

TABLE 5.8 *(Continued)*

Sl. No.	Botanical Name	Family	Vernacular Name	Ethnic Community	Plant part used	Mode of use
41.	*Shorea roxburghii* G Don	Dipterocarpaceae	Kungilium	Palliyar Pulayan	Cotyledons	Cotyledons are taken as food in boiled form
42.	*Sterculia foetida* L.	Sterculiaceae	Kongatti	Palliyar Pulayan	Kernel	Raw/Boiled/Roasted
43.	*Sterculia guttata* Roxb. ex DC.	Sterculiaceae	Kattuiluppai	Palliyar	Kernel	Raw/Boiled/Roasted
44.	*Sterculia urens* Roxb.	Sterculiaceae	Vennaali	Palliyar Pulayan	Kernel	Raw/Boiled
45.	*Strychnos nux-vomica* L.	Loganiaceae	Theathankottai	Palliyar Pulayan Kanikkar	Kernel	Raw/Roasted
46.	*Tamarindus indica* L.	Caesalpiniaceae	Puli	Palliyar Pulayan Kanikkar Valaiyan	Kernel	Raw/Boiled/Roasted
47.	*Teramnus labialis* (L.f) Spreng.	Fabaceae	Kattukanam	Palliyar Pulayan Kanikkar Valaiyan	Seed	Raw/Boiled/Roasted
48.	*Terminalia bellirica* (Gaertn.) Roxb.	Combretaceae	Tani	Palliyar Kanikkar	Kernel	Raw/Roasted
49.	*Terminalia chebula* Retz.	Combretaceae	Kadukkai	Palliyar Pulayan Kanikkar Valaiyan	Kernel	Soaked, dried and powdered kernels used in the preparation of 'dosai' and 'kali'
50.	*Vigna bourneae* Gamble	Fabaceae	Kattu payaru	Palliyar Pulayan	Seed	Raw/Boiled/Roasted
51.	*Vigna radiata* (L.) Wilczek var. *sublobata* (Roxb.) Verdc.	Fabaceae	Kattupasipayaru	Palliyar Pulayan Valaiyan	Seed	Raw/Boiled/Roasted

TABLE 5.8 *(Continued)*

Sl. No.	Botanical Name	Family	Vernacular Name	Ethnic Community	Plant part used	Mode of use
52.	*Vigna trilobata* (L.) Verdc.	Fabaceae	Kalpayaru	Palliyar Pulayan Valaiyan	Seed	Raw/Boiled/Roasted
53.	*Vigna unguiculata* (L.) Walp. subsp. *cylindrica* (L) Eselt.	Fabaceae	Panni Minnppayaru	Palliyar Pulayan Valaiyan	Seed	Raw/Boiled/Roasted
54.	*Vigna unguiculata* (L.) Walp. subsp. *unguiculata*.	Fabaceae	Kattukanam	Palliyar Pulayan	Seed	Raw/Boiled/Roasted
55.	*Ziziphus rugosa* Lam	Rhamnaceae	Mullakotti	Palliyar Pulayan	Kernel	Kernel is taken as food in roasted form or in idly natural state.
56.	*Ziziphus xylopyrus* (Retz.) Willd.	Rhamnaceae	Mullukottai	Palliyar Kanikkar Valaiyan	Kernel	Raw/Roasted
57.	*Xylia xylocarpa* (Roxb.) Taub.	Mimosaceae	Iral	Palliyar	Kernel	Roasted

5.1.2 PROXIMATE COMPOSITION

5.1.2.1 CRUDE PROTEIN

Proximate composition (Tables 5.9–5.12) reveals that the apical meristem of *Phoenix pusilla, kernels* of *Moringa concanensis*, *Sterculia urens*, *Tamarindus indica*, *Terminalia bellirica,* seeds of *Mucuna pruriens* var *pruriens*, *M. pruriens* var *utilis* (back colored seed coat) and *M. pruriens* var *utilis* (white colored seed coat) contain more than 25% crude protein.

5.1.2.2 CRUDE LIPID (ESTER EXTRACT)

The pith of *Arenga wightii*, tubers of *Asparagus racemosus*, *Dolichos trilobus*, rhizomes of *Curcuma neilgherrensis*, kernels of *Entada rheedi, Mucuna atropurpurea, Sterculia guttata, S. urens, Ziziphus rugosa, Z. xylopyrus* and the seed of *Ocimum gratissimum* and *Sesamum indicum* exhibit more than 10% of crude lipid content. Wild edible seed kernels of *Canarium strictum*, *Elaeocarpus tectorius, Givotia rottleriformis, Moringa concanensis, Sterculia foetida, Terminalia bellirica,* and *T. chebula* contain more than 30% of crude lipid content.

5.1.2.3 TOTAL DIETARY FIBER (TDF)

Total dietary fiber content was found to be relatively high (above 10%) in tuber of *Curculigo orchioides*, the rhizome of *Curcuma neilgherrensis,* the root of *Ipomoea staphylina* and kernel of *Tamarindus indica*.

5.1.2.4 ASH

The ash content ranged between 0.71% and 9.86% in different plant parts of the various plant species currently studied. The tuber of *Cyphostemma setosum* exhibits around 9% of ash content.

5.1.2.5 NFE (NITROGEN FREE EXTRACTIVES) OR TOTAL CRUDE CARBOHYDRATES

The content of the total crude carbohydrates of all the investigated samples ranged between 11.60% and 94.66%. The NFE values of piths of *Arenga*

TABLE 5.9 Proximate Composition of Edible Piths and Apical Meristem (g 100^{-1})[a]

Sl. No.	Botanical Name Parts Used	Moisture	Crude Protein (Kjeldahl N × 6.25)	Crude lipid	Total Dietary Fiber (TDF)	Ash	Nitrogen Free Extractive (NFE)	Calorific value ($kJ100g^{-1}DM$)
1.	*Arenga wightii* (Pith)	33.84 ± 0.15	1.75 ± 0.14	12.51 ± 0.18	1.23 ± 0.11	0.71 ± 0.31	83.80	1900.31
2.	*Caryota urens* (Pith)	12.56 ± 0.14	3.50 ± 0.14	6.36 ± 0.26	1.63 ± 0.04	6.56 ± 0.25	81.95	1666.79
3.	*Phoenix pusilla* (Apical meristem)	5.10 ± 0.38	29.50 ± 0.14	9.87 ± 0.48	9.69 ± 0.35	8.58 ± 0.18	42.36	1572.16

[a] All the values are means of triplicate determinations expressed on dry weight basis.

[±] Denotes standard error.

TABLE 5.10 Proximate composition of Edible Tubers, Rhizomes, Corms and Root-Types (g $100g^{-1}$)[a]

Sl. No.	Botanical Name	Moisture	Crude Protein (Kjeldhal N × 6.25)	Crude Lipid	Total Dietary Fibre (TDF)	Ash	Nitrogen Free Extractive (NFE)	Calorific value (KJ $100g^{-1}$ DM)
1.	*Abelmoschus moschatus* (Root)	65.16 ± 0.15	5.35 ± 0.025	6.12 ± 0.20	9.08 ± 0.14	4.80 ± 0.29	74.65	1566.72
2.	*Amorphophallus paeoniifolius* var. campanulatus (Corm)	80.00 ± 0.28	8.65 ± 0.12	5.35 ± 0.04	6.25 ± 0.14	3.18 ± 0.02	76.47	1624.87
3.	*Amorphophallus sylvaticus* (Corm)	67.89 ± 0.17	7.09 ± 0.40	6.35 ± 0.16	7.24 ± 0.21	3.12 ± 0.44	76.20	1630.34
4.	*Aponogeton natans* (Tubers)	9.28 ± 0.18	5.25 ± 0.04	2.90 ± 0.01	3.92 ± 0.02	2.84 ± 0.03	85.09	1618.00
5.	*Alocasia macrorhiza* (Tubers)	89.04 ± 2.06	3.47 ± 0.03	2.38 ± 0.03	3.45 ± 0.03	2.16 ± 0.01	88.54	1626.29
6.	*Argyreia pilosa* (Root)	72.48 ± 0.36	7.00 ± 0.50	3.85 ± 0.16	1.48 ± 0.05	2.83 ± 0.23	84.84	1678.87
7.	*Asparagus racemosus* (Tubers)	78.39 ± 0.31	6.73 ± 0.18	10.32 ± 0.18	7.35 ± 0.40	4.89 ± 0.17	70.71	1682.31
8.	*Borassus flabellifer* (Root)	62.48 ± 1.46	4.56 ± 0.03	3.31 ± 0.04	5.28 ± 0.06	3.28 ± 0.04	83.57	1596.56
9.	*Canna indica* (Rhizome)	51.60 ± 0.31	9.62 ± 0.27	5.30 ± 0.19	9.20 ± 0.10	4.01 ± 0.03	71.78	1559.19
10.	*Cissus vitiginea* (Tubers)	87.65 ± 2.65	3.93 ± 0.16	2.24 ± 0.04	4.48 ± 0.08	7.14 ± 0.02	82.20	1522.98
11.	*Colocasia esculenta* (Corm)	63.77 ± 0.10	1.63 ± 0.07	8.42 ± 0.24	7.92 ± 0.35	4.54 ± 0.15	77.49	1638.31
12.	*Costus speciosus* (Rhizome)	55.86 ± 1.14	4.86 ± 0.13	3.31 ± 0.06	3.86 ± 0.04	4.24 ± 0.10	83.73	1604.24
13.	*Curculigo orchioides* (Tubers)	67.44 ± 0.23	9.54 ± 0.44	4.44 ± 0.27	10.22 ± 0.12	3.53 ± 0.19	72.27	1533.62
14.	*Curcuma neilgherrensis* (Rhizome)	66.23 ± 0.28	19.00 ± 0.14	11.35 ± 0.24	10.32 ± 0.11	5.13 ± 0.09	54.20	1650.34
15.	*Cycas circinalis* (Tubers)	63.47 ± 1.86	9.18 ± 0.11	4.09 ± 0.11	3.93 ± 0.02	4.29 ± 0.12	78.50	1618.61
16.	*Cyphostemma setosum* (Tubers)	93.02 ± 3.56	4.37 ± 0.11	5.77 ± 0.03	3.34 ± 0.02	9.86 ± 0.02	76.65	1570.73
17.	*Decalepis hamiltonii* (Tubers)	78.73 ± 1.52	4.37 ± 0.09	10.24 ± 0.22	4.15 ± 0.03	8.70 ± 0.02	72.53	1670.44
18.	*Dioscorea alata* (Tubers)	82.91 ± 0.41	7.57 ± 0.11	5.28 ± 0.18	3.99 ± 0.11	3.56 ± 0.02	79.63	1655.30

TABLE 5.10 *(Continued)*

Sl. No.	Botanical Name	Moisture	Crude Protein (Kjeldhal N × 6.25)	Crude Lipid	Total Dietary Fibre (TDF)	Ash	Nitrogen Free Extractive (NFE)	Calorific value (KJ $100g^{-1}$ DM)
19.	*Dioscorea bulbifera* var. *vera* (Tubers)	68.70 ± 0.25	5.16 ± 0.23	9.13 ± 0.18	1.23 ± 0.06	2.91 ± 0.30	81.57	1792.59
20.	*Dioscorea esculenta* (Tubers)	83.37 ± 0.25	9.76 ± 0.23	4.68 ± 0.14	6.62 ± 0.21	5.17 ± 0.05	73.77	1571.39
21.	*Dioscorea hispida* (Tubers)	72.26 ± 2.16	6.88 ± 0.15	4.28 ± 0.08	5.26 ± 0.11	4.88 ± 0.12	78.70	1590.54
22.	*Dioscorea oppositifolia* var. *dukhumensis* (Tubers)	81.90 ± 0.18	13.80 ± 0.28	6.33 ± 0.34	3.92 ± 0.02	1.60 ± 0.17	74.35	1710.75
23.	*Dioscorea oppositifolia* var. *oppositifolia* (Tubers)	69.03 ± 0.51	6.31 ± 0.35	2.51 ± 0.21	8.97 ± 0.04	6.39 ± 0.26	75.82	1466.20
24.	*Dioscorea pentaphylla* . var *pentaphylla* (Tubers)	73.46 ± 0.27	5.38 ± 0.13	6.01 ± 0.45	7.04 ± 0.07	1.58 ± 0.11	79.99	1652.26
25.	*Dioscorea spicata* (Tubers)	89.26 ± 3.08	6.38 ± 0.08	4.78 ± 0.12	4.67 ± 0.03	5.18 ± 0.01	78.99	1605.89
26.	*Dioscorea tomentosa* (Tubers)	71.86 ± 0.47	8.51 ± 0.27	5.88 ± 0.19	2.24 ± 0.14	2.54 ± 0.39	80.83	1713.71
27.	*Dioscorea wallichii* (Tubers)	76.36 ± 0.27	10.76 ± 0.18	3.34 ± 0.04	7.48 ± 0.13	6.36 ± 0.05	72.06	1509.01
28.	*Dolichos trilobus* (Tubers)	72.38 ± 0.27	7.08 ± 0.36	10.80 ± 0.21	3.16 ± 0.10	3.02 ± 0.28	75.94	1793.60
29.	*Hemidesmus indicus* var. *indicus* (Root)	19.40 ± 0.78	4.37 ± 0.11	6.17 ± 0.07	3.15 ± 0.03	3.00 ± 0.12	83.30	1696.86
30.	*Hemidesmus indicus* var. *pubescens* (Root)	24.38 ± 0.52	5.28 ± 0.09	4.21 ± 0.06	3.25 ± 0.02	2.86 ± 0.08	84.40	1656.37
31.	*Ipomoea staphylina* (Root)	65.52 ± 0.19	4.41 ± 0.17	2.39 ± 0.09	10.25 ± 0.19	4.80 ± 0.29	78.15	1468.36
32.	*Kedrostis foetidissima* (Tubers)	80.76 ± 2.14	11.37 ± 0.03	5.09 ± 0.03	5.58 ± 0.06	7.36 ± 0.13	70.59	1560.79
33.	*Maerua oblongifolia* (Tubers)	55.48 ± 1.12	6.68 ± 0.06	4.21 ± 0.02	4.86 ± 0.05	3.38 ± 0.11	80.87	1620.80

TABLE 5.10 *(Continued)*

Sl. No.	Botanical Name	Moisture	Crude Protein (Kjeldhal N × 6.25)	Crude Lipid	Total Dietary Fibre (TDF)	Ash	Nitrogen Free Extractive (NFE)	Calorific value (KJ 100g^{-1} DM)
34.	*Manihot esculenta* (Tubers)	65.95 ± 1.26	3.50 ± 0.05	2.44 ± 0.03	2.47 ± 0.05	4.31 ± 0.14	87.28	1608.01
35.	*Maranta arundinacea* (Rhizome)	78.08 ± 2.46	13.13 ± 0.06	1.12 ± 0.01	3.48 ± 0.11	2.10 ± 0.01	77.17	1550.23
36.	*Momordica dioica* (Tubers)	77.78 ± 3.10	3.50 ± 0.03	5.08 ± 0.03	4.14 ± 0.50	5.36 ± 0.01	81.92	1618.03
37.	*Nephrolepis auriculata* (Tubers)	90.40 ± 1.22	8.75 ± 0.04	2.26 ± 0.01	2.13 ± 0.02	1.90 ± 0.01	84.96	1650.15
38.	*Nymphaea pubescens* (Tubers)	87.55 ± 3.66	9.62 ± 0.11	2.95 ± 0.01	3.15 ± 0.11	6.95 ± 0.04	77.29	1563.91
39.	*Nymphaea rubra* (Tubers)	81.33 ± 2.08	8.31 ± 0.12	5.05 ± 0.03	3.74 ± 0.09	3.80 ± 0.03	79.09	1650.13
40.	*Parthenocissus neilgherriensis* (Tubers)	79.29 ± 3.24	5.25 ± 0.12	4.89 ± 0.01	3.44 ± 0.03	8.96 ± 0.11	77.46	1565.61
41.	*Sarcostemma acidum* (Tubers)	62.16 ± 2.26	4.24 ± 0.06	3.54 ± 0.03	3.33 ± 0.02	3.78 ± 0.07	89.11	1625.60
42.	*Sterculia urens* (Root)	74.51 ± 0.23	5.25 ± 0.14	7.40 ± 0.16	3.63 ± 0.25	4.05 ± 0.40	79.67	1697.14
43.	*Xanthosoma sagittifolium* (Corm)	69.33 ± 1.86	8.75 ± 0.21	7.42 ± 0.12	7.48 ± 0.03	4.53 ± 0.01	71.82	1625.25
44.	*Xanthosoma violaceum* (Corm)	90.75 ± 2.18	6.78 ± 0.13	3.42 ± 0.03	4.51 ± 0.06	3.46 ± 0.03	81.83	1608.72

[a] All the values are means of triplicate determinations expresses on dry weight basis.

[±] Denotes standard error.

TABLE 5.11 Proximate Composition of Edible Greens (Leaves) (g $100g^{-1}$)[a]

Sl. No	Name of the Plant	Moisture	Crude Protein (Kjeldahl N × 6.25)	Crude lipid	Total Dietary Fiber (TDF)	Ash	Nitrogen Free Extractive (NFE)	Calorific value (kJ 100 g^{-1}DM)
1.	*Acacia caesia*	70.19 ± 0.36	4.37 ± 0.09	1.77 ± 0.01	2.04 ± 0.03	1.55 ± 0.13	90.27	1647.22
2.	*Acacia grahamii*	64.32 ± 0.82	2.86 ±	1.58 ± 0.02	0.26 ± 0.02	1.86 ± 0.02	91.44	1634.38
3.	*Achyranthes aspera*	66.52 ± 0.56	2.56 ± 0.06	1.38 ± 0.02	2.56 ± 0.02	1.94 ± 0.06	91.56	1623.83
4.	*Achyranthes bidentata*	68.47 ± 0.21	1.75 ± 0.01	1.14 ± 0.01	2.12 ± 0.06	1.39 ± 0.01	93.60	1635.32
5.	*Allmania nodiflora* var. angustifolia	72.36 ± 0.61	3.36 ± 0.05	2.24 ± 0.02	2.76 ± 0.04	2.01 ± 0.02	89.63	1637.38
6.	*Allmania nodiflora* var. procumbens.	70.86 ± 0.54	3.86 ± 0.07	2.38 ± 0.04	2.04 ± 0.03	2.21 ± 0.03	89.51	1649.01
7.	*Aloe vera*	73.14 ± 0.78	4.58 ± 0.09	2.96 ± 0.06	2.56 ± 0.04	2.36 ± 0.05	87.54	1649.99
8.	*Alternanthera bettzickiana*	71.24 ± 0.68	4.14 ± 0.06	2.34 ± 0.02	2.36 ± 0.03	2.88 ± 0.06	88.28	1658.63
9.	*Alternanthera sessilis*	72.80 ± 0.86	4.76 ± 0.08	1.86 ± 0.03	2.14 ± 0.02	2.56 ± 0.03	88.68	1630.57
10.	*Amaranthus roxburghianus*	72.16 ± 0.56	2.68 ± 0.02	0.92 ± 0.01	2.16 ± 0.02	2.32 0.02	91.92	1614.50
11.	*Amaranthus spinosus*	71.23 ± 0.38	1.75 ± 0.01	1.06 ± 0.02	2.84 ± 0.03	1.67 ± 0.02	92.68	1616.94
12.	*Amaranthus tricolor*	72.50 ± 0.72	3.14 ± 0.06	1.24 ± 0.06	2.40 ± 0.04	2.48 ± 0.03	90.74	1614.54
13.	*Amaranthus viridis.*	70.84 ± 0.84	3.36 ± 0.08	1.01 ± 0.01	2.42 ± 0.03	2.30 ± 0.03	88.06	1628.76
14.	*Amorphophallus sylvaticus*	68.33 ± 0.56	4.82 ± 0.06	2.06 ± 0.02	2.22 ± 0.04	2.84 ± 0.03	88.06	1628.76
15.	*Asystasia gangetica*	71.65 ± 0.12	3.50 ± 0.03	1.03 ± 0.01	2.13 ± 0.01	1.79 ± 0.03	91.53	1625.83
16.	*Basella alba var.alba*	73.40 ± 0.77	13.12 ± 0.06	1.89 ± 0.02	3.75 ± 0.07	2.31 ± 0.07	85.93	1608.48
17.	*Begonia malabarica*	69.39 ± 0.66	2.12 ± 0.03	1.08 ± 0.03	2.86 ± 0.03	2.59 ± 0.04	91.35	1601.67
18.	*Borahavia diffusa*	71.88 ± 0.54	2.10 ± 0.01	1.03 ± 0.02	2.34 ± 0.06	2.51 ± 0.03	92.02	1610.63
19.	*Boerhavia erecta*	73.26 ± 0.72	3.42 ± 0.03	1.18 ± 0.03	2.50 ± 0.03	2.66 ± 0.04	90.24	1608.61
20.	*Borassus flabellifer*	71.30 ± 0.82	2.26 ± 0.02	0.94 ± 0.04	2.18 ± 0.06	2.72 ± 0.06	91.90	1607.91
21.	*Brassica juncea*	70.33 ± 0.03	7.87 ± 0.09	1.90 ± 0.01	3.21 ± 0.08	1.58 ± 0.04	85.44	1629.90

TABLE 5.11 *(Continued)*

Sl. No	Name of the Plant	Moisture	Crude Protein (Kjeldahl N × 6.25)	Crude lipid	Total Dietary Fiber (TDF)	Ash	Nitrogen Free Extractive (NFE)	Calorific value (kJ 100 g^{-1}DM)
22.	*Canthium parvifolium*	72.66 ± 0.79	3.18 ± 0.03	1.24 ± 0.02	2.28 ± 0.04	2.12 ± 0.03	91.18	1622.56
23.	*Cassia obtusifolia*	71.06 ± 0.56	3.38 ± 0.03	1.06 ± 0.01	2.26 ± 0.03	1.58 ± 0.04	91.72	1628.13
24.	*Cassia tora*	70.38 ± 0.34	3.56 ± 0.04	1.24 ± 0.01	1.94 ± 0.04	2.06 ± 0.03	91.20	1629.24
25.	*Capsicum annuum*	72.28 ± 0.18	1.80 ± 0.01	1.21 ± 0.08	1.89 ± 0.11	1.65 ± 0.03	93.45	1636.29
26.	*Capsicum frutescens*	72.62 ± 0.31	2.12 ± 0.03	0.84 ± 0.02	1.86 ± 0.03	2.03 ± 0.02	93.15	1622.68
27.	*Cardiospermum canescens*	69.31 ± 0.26	2.84 ± 0.02	0.92 ± 0.02	2.04 ± 0.02	1.96 ± 0.03	92.24	1622.52
28.	*Cardiospermum helicacabum*	68.48 ± 0.33	3.11 ± 0.02	1.21 ± 0.01	2.34 ± 0.03	1.66 ± 0.02	91.68	1628.61
29.	*Cardiospermum microcarpa*	70.30 ± 0.46	2.24 ± 0.03	0.86 ± 0.02	1.56 ± 0.02	1.33 ± 0.03	94.01	1639.79
30	*Celosia argentea*	69.39 ± 0.30	3.36 ± 0.03	1.36 ± 0.03	2.06 ± 0.03	1.82 ± 0.06	91.40	1633.76
31	*Centella asiatica*	70.18 ± 0.18	3.56 ± 0.02	1.52 ± 0.03	1.33 ± 0.02	2.20 ± 0.04	91.39	1642.97
32.	*Cissus quadrangularis*	73.56 ± 0.62	2.84 ± 0.02	1.01 ± 0.01	1.52 ± 0.03	1.56 ± 0.03	93.07	1639.77
33.	*Cissus vitiginea*	72.82 ± 0.56	2.66 ± 0.02	0.96 ± 0.02	1.30 ± 0.02	1.14 ± 0.02	93.94	1649.41
34.	*Cleome gynandra*	70.32 ± 0.26	1.82 ± 0.01	0.54 ± 0.01	1.72 ± 0.01	1.26 ± 0.01	94.66	1631.57
35	*Cleome viscosa*	72.44 ± 0.56	3.16 ± 0.03	1.12 ± 0.01	2.22 ± 0.06	2.28 ± 0.06	91.22	1618.37
36.	*Coccinia grandis*	71.06 ± 0.48	3.33 ± 0.02	1.30 ± 0.01	2.26 ± 0.07	2.16 ± 0.04	90.95	1623.49
37.	*Cocculus hirsutus*	73.52 ± 0.42	2.98 ± 0.02	0.86 ± 0.02	1.78 ± 0.05	2.10 ± 0.03	92.28	1623.26
38.	*Colocasia esculenta*	72.32 ± 0.46	3.21 ± 0.04	1.18 ± 0.02	1.38 ± 0.04	2.66 ± 0.05	91.57	1627.31
39.	*Commelina benghalensis*	73.17 ± 0.17	4.38 ± 0.06	1.17 ± 0.12	3.36 ± 0.14	1.96 ± 0.08	89.13	1605.72
40.	*Commelina ensifolia*	74.12 ± 0.26	1.98 ± 0.05	0.91 ± 0.04	1.10 ± 0.08	1.36 ± 0.06	94.65	1648.03
41.	*Coriandrum sativum*	73.32 ± 0.23	2.21 ± 0.03	1.01 ± 0.03	1.08 ± 0.06	1.53 ± 0.05	94.17	1647.62
42.	*Cycas circinalis*	68.76 ± 0.18	2.16 ± 0.02	0.93 ± 0.02	1.14 ± 0.05	1.92 ± 0.04	93.85	1638.43

TABLE 5.11 *(Continued)*

Sl. No	Name of the Plant	Moisture	Crude Protein (Kjeldahl N × 6.25)	Crude lipid	Total Dietary Fiber (TDF)	Ash	Nitrogen Free Extractive (NFE)	Calorific value (kJ 100 g^{-1}DM)
43.	*Digera muricata*	69.32 ± 0.25	2.68 ± 0.03	1.01 ± 0.	1.36 ± 0.09	1.32 ± 0.03	93.63	1646.45
44.	*Diplocyclos palmatus*	76.55 ± 0.19	4.37 ± 0.05	1.33 ± 0.03	3.95 ± 0.13	2.11 ± 0.01	84.24	1529.92
45.	*Emilia sonchifolia*	70.82 ± 0.26	2.96 ± 0.03	0.96 ± 0.02	1.96 ± 0.07	2.01 ± 0.02	92.11	1623.86
46.	*Euphorbia hirta*	69.10 ± 0.22	2.09 0 ± 0.01	0.94 ± 0.03	1.11 ± 0.04	1.26 ± 0.02	94.60	1650.16
47.	*Gisekia pharnaceoides*	69.14 ± 0.18	3.16 ± 0.03	1.26 ± 0.02	2.12 ± 0.08	2.11 ± 0.03	91.35	1625.82
48.	*Glinus oppositifolius*	68.12 ± 0.36	2.98 ± 0.01	1.33 ± 0.01	2.22 ± 0.03	1.54 ± 0.03	91.93	1635.14
49.	*Heracleum rigens* var. rigens	71.66 ± 0.28	3.14 ± 0.02	1.26 ± 0.02	2.56 ± 0.02	1.82 ± 0.06	91.22	1623.31
50.	*Hybanthus enneaspermus*	70.18 ± 0.42	3.48 ± 0.06	1.54 ± 0.03	2.66 ± 0.01	1.92 ± 0.08	90.40	1625.85
51.	*Ipomoea aquatic*	69.76 ± 0.32	2.56 ± 0.03	1.41 ± 0.06	2.44 ± 0.03	1.40 ± 0.10	92.19	1635.48
52.	*Ipomoea pes-tigridis*	70.14 ± 0.21	2.76 ± 0.06	1.12 ± 0.02	2.33 ± 0.22	1.39 ± 0.05	92.40	1631.40
53.	*Jasminum auriculatum*	64.33 ± 0.33	2.86 ± 0.04	1.31 ± 0.03	2.86 ± 0.02	2.01 ± 0.04	90.96	1629.21
54.	*Jasminum calophyllum*	67.28 ± 0.41	2.54 ± 0.02	1.44 ± 0.04	2.14 ± 0.06	2.14 ± 0.06	91.74	1628.76
55.	*Kalanchoe pinnata*	74.36 ± 0.38	3.26 ± 0.08	1.39 ± 0.03	2.18 ± 0.04	2.36 ± 0.03	90.81	1623.37
56.	*Lathyrus sativus*	67.30 ± 0.26	61.14 ± 0.14	1.28 ± 0.02	2.26 ± 0.06	1.88 ± 0.02	88.44	1627.74
57.	*Leucas montana* var. wightii.	69.03 ± 0.41	5.25 ± 0.11	1.90 ± 0.10	2.27 ± 0.03	1.85 ± 0.13	88.73	1641.09
58.	*Mollugo pentaphylla*	72.16 ± 0.43	4.32 ± 0.09	1.56 ± 0.09	2.28 ± 0.04	2.16 ± 0.01	89.68	1628.62
59.	*Moringa concanensis*	65.37 ± 0.28	6.76 ± 0.08	1.78 ± 0.11	2.52 ± 0.05	2.30 ± 0.08	86.64	1626.89
60.	*Mukia maderaspatana*	72.41 ± 0.19	4.37 ± 0.13	1.58 ± 0.11	2.73 ± 0.01	2.32 ± 0.09	89.00	1618.84
61.	*Murraya koenigii*	71.04 ± 0.33	3.38 ± 0.11	1.20 ± 0.09	2.66 ± 0.02	1.98 ± 0.03	90.78	1617.71
62.	*Murraya paniculata*	64.96 ± 0.28	3.18 ± 0.06	1.36 ± 0.06	2.44 ± 0.04	1.68 ± 0.03	91.34	1622.97
63.	*Oxalis corniculata*	77.33 ± 0.41	9.62 ±0 .18	1.87 ± 0.03	2.15 ±0.06	1.95 ± 0.10	84.41	1640.80

TABLE 5.11 *(Continued)*

Sl. No	Name of the Plant	Moisture	Crude Protein (Kjeldahl N × 6.25)	Crude lipid	Total Dietary Fiber (TDF)	Ash	Nitrogen Free Extractive (NFE)	Calorific value (kJ 100 g^{-1}DM)
64.	*Oxalis latifolia*	76.30 ± 0.24	7.36 ± 0.04	1.57 ± 0.08	2.05 ± 0.03	2.06 ± 0.06	86.96	1634.33
65.	*Peperomia pellucida*	73.18 ± 0.36	4.21 ± 0.03	1.48 ± 0.06	1.98 ± 0.06	2.14 ± 0.05	90.19	1632.28
66.	*Physalis minima* var. indica	70.56 ± 0.32	3.68 ± 0.04	1.28 ± 0.04	2.84 ± 0.08	1.86 ± 0.04	90.34	1618.39
67.	*Portulaca oleracea* var. oleracea	69.32 ± 0.26	3.44 ± 0.06	1.68 ± 0.05	2.16 ± 0.06	2.12 ± 0.07	90.60	1633.80
68.	*Portulaca quadrifida*	68.56 ± 0.23	3.33 ± 0.05	1.54 ± 0.06	2.53 ± 0.05	2.04 ± 0.08	90.56	1626.02
69.	*Premna corymbosa*	69.76 ± 0.22	3.28 ± 0.04	1.33 ± 0.03	2.41 ± 0.06	2.16 ± 0.06	90.82	1621.61
70.	*Psilanthus wightianus*	72.14 ± 0.11	2.98 ± 0.03	1.62 ± 0.04	2.33 ± 0.08	2.03 ± 0.04	91.04	1631.21
71.	*Sarcostemma acidum*	74.30 ± 0.29	3.36 ± 0.03	1.56 ± 0.08	2.68 ± 0.07	2.16 ± 0.03	90.24	1622.69
72.	*Sesbania grandiflora*	63.16 ± 0.21	5.30 ± 0.11	2.12 ± 0.03	2.98 ± 0.04	1.38 ± 0.01	88.22	1641.71
73.	*Solanum anguivi* var. multiflora	60.14 ± 0.18	4.16 ± 0.09	2.02 ± 0.03	2.66 ± 0.03	1.58 ± 0.03	89.58	1641.61
74.	*Solanum nigrum*	70.21 ± 0.19	5.25 ± 0.06	1.70 ± 0.02	3.12 ± 0.18	1.15 ± 0.04	88.78	1634.39
75.	*Solanum trilobatum*	64.36 ± 0.26	3.38 ± 0.05	1.68 ± 0.01	2.92 ± 0.14	1.62 ± 0.05	90.40	1630.22
76.	*Tamarindus indica*	64.40 ± 0.22	3.78 ± 0.04	1.54 ± 0.01	2.54 ± 0.11	1.38 ± 0.03	90.76	1636.88
77.	*Tinospora cordifolia*	66.36 ± 0.24	2.98 ± 0.03	1.86 ± 0.02	2.78 ± 0.09	1.32 ± 0.02	91.06	1640.59
78.	*Trianthema portulacastrum*	68.38 ± 0.11	3.36 ± 0.04	1.98 ± 0.01	2.88 ± 0.15	1.18 ± 0.01	90.60	1643.78
79.	*Vigna radiata*	64.36 ± 0.34	4.48 ± 0.05	2.21 ± 0.01	2.06 ± 0.09	1.36 ± 0.02	89.89	1659.30
80.	*Vigna trilobata*	65.32 ± 0.33	4.56 ± 0.04	2.06 ± 0.03	2.14 ± 0.07	1.52 ± 0.03	89.72	1652.14

[a] All the values are means of triplicate determinations expressed on dry weight basis.

± Denotes standard error.

TABLE 5.12 Proximate Composition of Wild Edible Seeds and Seed Components (g $100g^{-1}$) [a]

Sl. No	Name of the Plant	Components						
		Moisture	**Crude Protein (Kjeldahl N × 6.25)**	**Crude lipid**	**Total Dietary Fiber (TDF)**	**Ash**	**Nitrogen Free Extractive (NFE)**	**Calorific value (kJ100g^{-1} DM)**
1.	*Artocarpus heterophyllus* (Kernel)	12.38 ± 0.38	8.34 ± 0.11	4.26 ± 0.06	6.66 ± 0.14	4.88 ± 0.12	75.86	1566.74
2.	*Atylosia scarabaeoides* (Seed)	8.34 ± 0.09	17.33 ± 0.14	4.56 ± 0.12	7.21 ± 0.31	3.18 ± 0.11	67.72	1592.25
3.	*Bambusa arundinacea* (Seed)	7.36 ± 0.11	5.36 ± 0.08	4.32 ± 0.03	4.56 ± 0.12	4.16 ± 0.06	81.60	1615.10
4.	*Borassus flabellifer* (Endosperm)	13.86 ± 0.28	5.66 ± 0.16	4.16 ± 0.07	3.38 ± 0.04	3.98 ± 0.05	82.82	1634.45
5.	*Bupleurum wightii*.var. *ramosissimum* (Seed)	8.56 ± 0.08	6.24 ± 0.04	4.08 ± 0.06	3.86 ± 0.05	3.30 ± 0.04	82.52	1636.11
6.	*Canarium strictum* (Kernel)	7.39 ± 0.23	11.12 ± 0.73	55.53 ± 0.19	1.31 ± 0.05	6.03 ± 0.14	26.01	2713.55
7.	*Canavalia gladiata* (Seed)	7.60 ± 0.78	12.93 ± 0.14	9.31 ± 0.55	3.28 ± 0.12	4.03 ± 0.04	70.45	1743.10
8.	*Canavalia virosa* (Seed)	10.40 ± 0.07	17.50 ± 0.11	3.10 ± 0.03	3.24 ± 0.01	3.10 ± 0.01	73.06	1629.92
9.	*Capparis zeylanica* (Kernel)	7.35 ± 0.08	10.32 ± 0.08	5.36 ± 0.02	4.36 ± 0.05	4.11 ± 0.05	75.85	1641.11
10.	*Celtis philippensis* var. *wightii* (Seed)	6.88 ± 0.07	9.18 ± 0.14	5.66 ± 0.03	4.18 ± 0.04	4.38 ± 0.07	76.60	1645.91
11.	*Chamaecrista absus* (Seed)	8.98 ± 0.18	16.38 ± 0.36	4.28 ± 0.18	4.66 ± 0.32	3.88 ± 0.05	70.80	1617.26
12.	*Cycas circinalis* (Kernel)	8.48 ± 0.31	14.70 ± 0.44	2.43 ± 0.25	4.01 ± 0.11	2.83 ± 0.05	76.03	1606.80
13.	*Dolichos lablab* var. *vulgaris* (seed)	8.21 ± 0.11	18.61 ± 0.32	4.38 ± 0.16	5.24 ± 0.14	3.78 ± 0.06	67.99	1161.35
14.	*Dolichos trilobus* (Seed)	7.89 ± 0.12	17.81 ± 0.05	6.34 ± 0.12	6.56 ± 0.11	4.88 ± 0.11	64.41	1612.09
15.	*Drypetes sepiaria* (Seed)	7.56 ± 0.11	10.36 ± 0.18	5.32 ± 0.06	5.66 ± 0.04	4.88 ± 0.07	73.78	1605.70
16.	*Elaeocarpus tectorius* (Kernel)	5.34 ± 0.09	17.87 ± 0.22	41.54 ± 0.17	1.82 ± 0.04	3.13 ± 0.08	35.64	2459.68
17.	*Eleusine coracana* (Seed)	5.38 ± 0.14	9.54 ± 0.06	4.38 ± 0.05	4.12 ± 0.03	4.66 ± 0.05	77.30	1615.35
18.	*Ensete superbum* (Seed)	6.76 ± 0.11	9.86 ± 0.04	3.31 ± 0.04	4.11 ± 0.06	4.86 ± 0.04	77.86	1589.71

TABLE 5.12 *(Continued)*

Sl. No	Name of the Plant	Components						
		Moisture	Crude Protein (Kjeldahl N × 6.25)	Crude lipid	Total Dietary Fiber (TDF)	Ash	Nitrogen Free Extractive (NFE)	Calorific value ($kJ100g^{-1}$ DM)
19.	*Entada rheedi* (Seed)	4.53 ± 0.37	17.01 ± 0.44	10.22 ± 0.21	9.37 ± 0.04	2.74 ± 0.03	60.66	1682.38
20.	*Givotia rottleriformis* (Kernel)	5.22 ± 0.08	14.58 ± 0.18	50.74 ± 0.24	6.42 ± 0.26	5.69 ± 0.13	22.61	2533.30
21.	*Heracleum rigens* var. *rigens* (Seed)	5.46 ± 0.07	9.11 ± 0.07	4.32 ± 0.08	4.10 ± 0.07	3.38 ± 0.01	79.09	1635.80
22.	*Impatiens balsamina* (Seed)	6.08 ± 0.06	8.32 ± 0.05	5.11 ± 0.06	4.12 ± 0.04	4.38 ± 0.03	78.07	1635.36
23.	*Lablab purpureus* var. *lignosus* (Seed)	11.26 ± 0.15	24.75 ± 0.82	8.33 ± 0.27	1.21 ± 0.06	3.99 ± 0.17	61.72	1758.09
24.	*Lablab purpureus* var. *purpureus* (Seed)	8.24 ± 0.08	22.68 ± 0.14	7.24 ± 0.08	3.06 ± 0.04	3.59 ± 0.26	63.43	1710.99
25.	*Macrotyloma uniflorum* (seed)	10.24 ± 0.14	18.26 ± 0.21	4.86 ± 0.04	5.24 ± 0.08	4.24 ± 0.07	67.60	1613.74
26.	*Moringa concanensis* (Kernel)	13.45 ± 0.27	43.00 ± 0.43	40.19 ± 0.45	1.36 ± 0.07	3.85 ± 0.06	11.60	2426.98
27.	*Mucuna atropurpurea* (Kernel)	9.56 ± 0.23	24.43 ± 0.29	13.92 ± 0.07	9.23 ± 0.13	2.58 ± 0.18	49.84	1765.09
28.	*Mucuna pruriens* var. *puriens* (Seed)	10.48 ± 0.17	29.36 ± 0.34	8.24 ± 0.06	6.54 ± 0.04	4.78 ± 0.03	59.32	1791.60
29.	*Mucuna pruriens* var. *utilis* (Black coloured seed coat) (Seed)	10.35 ± 0.08	28.75 ± 0.17	9.36 ± 0.05	7.68 ± 0.06	5.12 ± 0.01	49.09	1652.80
30	*Mucuna pruriens* var. *utilis.* (White coloured seed coat) (Seed)	11.25 ± 0.11	30.63 ± 0.14	8.74 ± 0.07	8.50 ± 0.05	4.12 ± 0.09	47.95	1641.78
31	*Neonotonia wightii.* var. *coimbatorensis* (Seed)	6.23 ± 0.07	15.13 ± 0.11	4.64 ± 0.14	9.48 ± 0.12	3.33 ± 0.08	67.42	1553.51
32.	*Ocimum gratissimum* (Seed)	9.16 ± 0.18	16.21 ± 0.25	20.31 ± 0.35	3.39 ± 0.09	4.75 ± 0.21	55.34	1960.57
33.	*Oryza meyeriana* var. *granulata* (Grain)	4.38 ± 0.06	10.12 ± 0.11	3.21 ± 0.06	3.38 ± 0.04	4.6 ± 0.07	79.23	1613.16

TABLE 5.12 *(Continued)*

Sl. No	Name of the Plant	Components						
		Moisture	Crude Protein (Kjeldahl N × 6.25)	Crude lipid	Total Dietary Fiber (TDF)	Ash	Nitrogen Free Extractive (NFE)	Calorific value (kJ100g^{-1} DM)
34.	*Pithecellobium dulce* (Aril)	12.37 ± 0.11	6.78 ± 0.07	2.36 ± 0.10	3.76 ± 0.14	3.86 ± 0.03	83.24	1592.31
35.	*Rhynchosia cana* (Seed)	10.49 ± 0.52	12.83 ± 0.36	3.32 ± 0.12	9.77 ± 0.07	2.44 ± 0.06	71.64	1535.31
36.	*Rhynchosia filipes* (Seed)	9.34 ± 0.46	16.92 ± 0.22	3.77 ± 0.28	8.72 ± 0.12	4.34 ± 0.09	66.25	1531.07
37.	*Rhynchosia rufescens* (Seed)	5.10 ± 0.01	19.40 ± 0.01	4.41 ± 0.01	8.44 ± 0.03	3.50 ± 0.02	64.25	1563.21
38.	*Rhynchosia suaveolens* (Seed)	5.14 ± 0.11	14.84 ± 0.07	3.16 ± 0.13	8.38 ± 0.17	4.11 ± 0.14	69.51	1527.78
39.	*Sapindus emarginatus* (Seed)	7.30 ± 0.06	11.36 ± 0.08	3.16 ± 0.04	3.35 ± 0.08	4.51 ± 0.06	77.62	1605.10
40.	*Sesamum indicum* (Seed)	2.94 ± 0.21	12.00 ± 0.14	20.57 ± 0.28	2.13 ± 0.06	5.98 ± 0.41	59.32	1966.53
41.	*Shorea roxburghii* (Cotyledon)	6.19 ± 0.05	10.98 ± 0.04	4.44 ± 0.06	4.36 ± 0.11	3.82 ± 0.05	76.40	1626.63
42.	*Sterculia foetida* (Kernel)	7.39 ± 0.29	12.00 ± 0.14	38.81 ± 0.41	2.48 ± 0.08	3.81 ± 0.02	42.90	2379.97
43.	*Sterculia guttata* (Kernel)	10.21 ± 0.12	5.00 ± 0.14	21.04 ± 0.32	1.44 ± 0.05	4.83 ± 0.02	67.69	2007.13
44.	*Sterculia urens*(Kernel)	14.56 ± 0.19	28.46 ± 0.36	23.00 ± 0.56	2.57 ± 0.01	3.77 ± 0.02	42.20	2047.12
45.	*Strychnos nux-vomica* (Kernel)	11.54 ± 0.22	4.56 ± 0.29	4.98 ± 0.14	5.23 ± 0.10	1.97 ± 0.08	83.26	1654.34
46.	*Tamarindus indica* (Kernel)	9.69 ± 0.23	26.22 ± 0.55	5.44 ± 0.02	10.34 ± 0.12	1.88 ± 0.04	56.10	1580.59
47.	*Teramnus labialis* (Seed)	11.64 ± 0.15	23.83 ± 0.44	4.42 ± 0.12	9.23 ± 0.06	4.89 ± 0.01	57.63	1527.02
48.	*Terminalia bellirica* (Kernel)	10.10 ± 0.21	27.13 ± 0.07	43.61 ± 0.26	1.49 ± 0.03	4.92 ± 0.06	22.85	2478.76
49.	*Terminalia chebula* (Kernel)	9.84 ± 0.16	22.18 ± 0.32	32.14 ± 1.12	2.98 ± 0.96	4.24 ± 0.32	38.46	2224.37
50.	*Vigna bourneae* (Seed)	6.94 ± 0.01	25.84 ± 0.56	5.80 ± 0.17	4.89 ± 0.05	3.94 ± 0.56	59.53	1644.34
51.	*Vigna radiata* var. *sublobata* (Seed)	5.34 ± 0.04	18.54 ± 0.07	5.34 ± 0.11	7.39 ± 0.18	4.56 ± 0.07	64.17	1582.58
52.	*Vigna trilobata* (Seed)	9.67 ± 0.21	20.22 ± 0.55	12.32 ± 0.26	6.14 ± 0.14	3.89 ± 0.03	57.43	1761.22

TABLE 5.12 *(Continued)*

Sl. No	Name of the Plant	Components						
		Moisture	Crude Protein (Kjeldahl N × 6.25)	Crude lipid	Total Dietary Fiber (TDF)	Ash	Nitrogen Free Extractive (NFE)	Calorific value ($kJ100g^{-1}$ DM)
53.	*Vigna unguiculata subsp. cylindrica* (Seed)	3.23 ± 0.03	13.96 ± 0.11	3.43 ± 0.08	3.36 ± 0.02	3.63 ± 0.03	75.62	1625.30
54.	*Vigna unguiculata* subsp. unguiculata (Seed)	9.73 ± 0.18	15.93 ± 0.46	4.55 ± 0.16	3.16 ± 0.03	3.18 ± 0.01	73.18	1659.67
55.	*Ziziphus rugosa* (Kernel)	8.34 ± 0.15	10.14 ± 0.38	12.36 ± 0.18	5.38 ± 0.46	3.08 ± 0.14	69.04	1788.28
56.	*Ziziphus xylopyrus* (Kernel)	9.23 ± 0.54	9.16 ± 0.26	15.80 ± 0.31	1.13 ± 0.06	2.66 ± 0.14	71.25	1938.51
57	*Xylia xylocarpa* (Kernel)	5.63 ± 0.11	17.36 ± 0.05	5.30 ± 0.14	4.46 ± 0.12	4.16 ± 0.08	68.72	1637.35

a All the values are means of triplicate determinations expressed on dry weight basis.

± Denotes standard error.

wightii, *Caryota urens*, tubers of *Aponogeton natans*, *Alocasia macrorhiza*, *Cissus vitiginea*, *Dioscorea bulbifera* var *vera*, *D.tomentosa*, *Maerua oblongifolia*, *Manihot esculenta*, *Nephrolepis auriculata*, *Sarcostemma acidum*, rhizome of *Costus speciosus*, corm of *Xanthosoma violaceum*, root of *Argyria pilosa*, *Borassus flabellifer*, *Hemidesmus indicus* var *pubescens*, and all the edible greens and seeds of *Bambusa arundinacea*, *Bupleuram wightii* var *ramosissimum* kernel of *Strychnous nux-vomica* and endosperm of *Borassus flabellifer* are found to exceed 80%.

5.1.2.6 CALORIFIC VALUE

The calorific value of the investigated seed kernels of *Canarium strictum*, *Elaeocarpus tectorius*, *Givotia rottleriformis*, *Moringa concanensis*, *Sterculia foetida*, *S. guttata*, *S. urens*, *Terminalia bellirica* and *T. chebula* is found to exceed 2000 KJ 100^{-1} DM.

5.1.3 MINERAL COMPOSITION

The data on the mineral profiles are furnished in Table 5.13–5.16 corms of *Colocasia esculenta*, *xanthosoma segittifolium*, leaf of *Basella alba*, kernels of *Entada rheedi*, *Mucuna atropurpurea*, *Tamarindus indica* and seeds of *Canavalia gladiata*, *Lablab purpureus* var_*lignosus*, *Rhynchosia filipes*, *Vigna bourneae*, *V. radiata* var. *sublobata* and *V.ungiculata* subsp *ungiculata* exhibited high level of potassium. The pith of *Caryota urens*, apical meristem of *Phoenix pusilla*, tubers of *Alocasia macrorhiza*, *Curculigo orchioides*, *Decalepis hamiltonii*, *Dioscorea pentaphylla* var. *pentaphylla*, *Dioscorea tomentosa* and corms of *Colocasia esculenta*, *Xanthosoma sagittifolium* and *X. violaceum* is found to be rich in sodium, potassium, calcium, iron, and copper.

5.1.3.1 TOTAL SOLUBLE CARBOHYDRATES, STARCH, AND TOTAL FREE AMINO ACIDS

The data on the total soluble carbohydrates, starch and total free amino acids of the investigated samples are preserved in Tables 5.17–5.20.

TABLE 5.13 Mineral Composition of Edible Piths and Apical Meristem (mg $100^{\neq 1}$)[a]

Sl. No.	Botanical Name	Sodium	Potassium	Calcium	Magnesium	Phosphorus	Zinc	Manganese	Iron	Copper
1.	*Arenga wightii* (Pith)	90.60 ± 0.08	868.70 ± 0.32	240.40 ± 0.04	248.15 ± 0.04	120.12 ± 0.03	3.30 ± 0.23	22.10 ± 0.12	3.50 ± 0.03	1.50 ± 0.10
2.	*Caryota urens* (Pith)	35.00 ± 0.15	748.10 ± 0.83	280.10 ± 0.31	1750.00 ± 1.12	134.50 ± 0.08	3.48 ± 0.10	2.30 ± 0.03	51.20 ± 0.14	1.80 ± 0.02
3.	*Phoenix pusilla* (Apical meristem)	15.30 ± 0.04	986.00 ± 0.85	110.30 ± 0.11	148.12 ± 0.07	140.00 ± 0.01	7.80 ± 0.04	6.50 ± 0.02	52.40 ± 0.18	1.40 ± 0.01

[a] All the values are means of triplicate determinations expressed on dry weight basis.

± Denotes standard error.

TABLE 5.14 Mineral Composition of Edible Tubers, Rhizomes, Corms and Root Types (mg 100^{-1})[a]

Sl. No.	Botanical Name	Sodium	Potassium	Calcium	Magnesium	Phosphorus	Iron	Zinc	Copper	Manganese
1.	*Abelmoschus moschatus* (Root)	47.50 ± 0.18	1138.00 ± 1.52	410.34 ± 0.31	144.00 ± 0.44	83.04 ± 0.31	40.10 ± 0.33	4.40 ± 0.02	1.60 ± 0.04	5.40 ± 0.11
2.	*Amorphophallus paeoniifolius* var. *campanulatus* (Corm)	38.30 ± 0.16	1248.16 ±1.32	660.32 ± 0.18	132.04 ± 0.21	115.36 ± 0.20	28.36 ± 0.15	5.32 ± 0.11	1.84 ± 0.01	5.56 ± 0.08
3.	*Amorphophallus sylvaticus* (Corm)	22.50 ± 0.12	1115.38 ± 1.32	520.00 ± 0.28	120.60 ± 0.12	124.40 ± 0.22	41.60 ± 1.31	7.88 ± 0.13	2.20 ± 0.02	6.00 ± 0.14
4.	*Aponogeton natans* (Tubers)	43.10 ± 0.10	1038.26 ± 1.24	312.30 ± 0.46	134.08 ± 0.26	68.05 ± 0.14	30.10 ± 0.44	2.20 ± 0.06	1.54 ± 0.01	3.34 ± 0.04
5.	*Alocasia macrorhiza* (Tubers)	196.01 ± 0.05	1941.36 ± 0.82	580.48 ± 0.16	154.10 ± 0.04	89.60 ± 0.12	72.74 ± 0.06	2.20 ± 0.02	7.26 ± 0.02	3.34 ± 0.03
6.	*Argyreia pilosa* (Root)	42.50 ± 0.14	838.41 ± 0.36	200.00 ± 0.08	264.54 ± 0.03	66.34 ± 0.07	23.24 ± 0.08	4.54 ± 0.18	2.60 ± 0.01	3.34 ± 0.03
7.	*Asparagus racemosus* (Tubers)	25.05 ± 0.32	548.00 ± 0.14	170.30 ± 0.14	280.10 ± 0.06	7942 ± 0.09	21.20 ± 0.03	2.06 ± 0.03	3.20 ± 0.01	13.80 ± 0.14
8.	*Borassus flabellifer* (Root)	16.36 ± 0.18	436.16 ± 0.22	186.32 ± 0.33	126.03 ± 0.04	82.14 ± 0.08	26.16 ± 0.06	2.28 ± 0.05	2.16 ± 0.03	8.36 ± 0.11
9.	*Canna indica* (Rhizome)	21.10 ± 0.28	979.16 ± 0.17	154.18 ± 0.12	134.12 ± 0.08	89.06 ± 0.07	11.53 ± 0.12	0.89 ± 0.04	3.36 ± 0.06	1.21 ± 0.03
10.	*Cissus vitiginea* (Tubers)	18.24 ± 0.11	638.36 ± 0.46	316.14 ± 0.03	88.30 ± 0.09	38.10 ± 0.18	26.08 ± 0.28	1.75 ± 0.01	2.54 ± 0.01	2.54 ± 0.01
11.	*Colocasia esculenta* (Corm)	200.10 ± 0.44	2041.35 ± 2.11	680.48 ± 0.07	200.48 ± 0.04	101.04 ± 1.21	82.74 ± 0.54	2.10 ± 0.11	9.94 ± 0.08	3.24 ± 0.08
12.	Costus speciosus	26.18 ± 0.09	786.52 ± 0.18	408.32 ± 0.08	280.12 ± 0.16	103.24 ± 0.52	32.26 ± 0.96	2.11 ± 0.06	1.06 ± 0.01	4.42 ± 0.07
13.	*Curculigo orchioides* (Tubers)	32.54 ± 0.08	668.00 ± 0.58	440.34 ± 0.11	360.30 ± 0.11	88.60 ± 0.86	124.38 ± 1.21	2.48 ± 0.03	2.34 ± 0.02	5.24 ± 0.01
14.	*Curcuma neilgherrensis* (Rhizome)	25.00 ± 0.21	980.40 ± 1.21	240.00 ± 0.18	420.34 ± 0.34	34.50 ± 0.16	35.40 ± 0.04	4.04 ± 0.06	2.78 ± 0.10	22.41 ± 0.28
15.	*Cycas circinalis* (Tubers)	94.10 ± 0.32	1548.36 ± 1.18	418.36 ± 0.04	138.10 ± 0.36	102.00 ± 0.91	31.00 ± 0.56	1.48 ± 0.01	5.24 ± 0.06	5.54 ± 0.13
16.	*Cyphostemma setosum* (Tubers)	54.30 ± 0.08	934.26 ± 0.18	346.14 ± 0.08	124.04 ± 0.56	44.68 ± 0.26	28.00 ± 0.33	1.40 ± 0.02	2.24 ± 0.01	3.16 ± 0.03
17.	*Decalepis hamiltonii* (Tubers)	64.00 ± 0.14	1134.08 ± 0.18	386.12 ± 0.26	104.08 ± 0.22	46.12 ± 0.08	51.10 ± 0.16	2.18 ± 0.03	1.34 ± 0.01	5.44 ± 0.04
18.	*Dioscorea alata* (Tubers)	44.56 ± 0.31	786.30 ± 0.14	448.36 ± 0.11	656.31 ± 0.07	140.14 ± 0.14	24.30 ± 0.19	2.26 ± 0.01	11.20 ± 0.14	6.36 ± 0.21
19.	*Dioscorea bulbifera* var. *vera* (Tubers)	66.78 ± 0.44	1600.31 ± 1.48	238.15 ± 0.09	411.17 ± 0.08	134.14 ± 0.53	4.90 ± 0.01	1.30 ± 0.01	2.74 ± 0.01	11.60 ± 0.17

TABLE 5.14 *(Continued)*

Sl. No.	Botanical Name	Sodium	Potassium	Calcium	Magnesium	Phosphorus	Iron	Zinc	Copper	Manganese
20.	*Dioscorea esculenta* (Tubers)	86.40 ± 0.14	1594.31 ± 1.34	314.01 ± 0.33	436.06 ± 0.54	138.10 ± 0.14	11.48 ± 0.11	1.76 ± 0.04	3.40 ± 0.01	5.46 ± 0.11
21.	*Dioscorea hispida*	56.32 ± 0.16	896.08 ± 0.62	206.31 ± 0.18	414.32 ± 0.33	101.36 ± 0.16	33.26 ± 0.11	2.48 ± 0.06	1.28 ± 0.01	5.26 ± 0.10
22.	*Dioscorea oppositifolia* var. *dukhumensis* (Tubers)	123.00 ± 0.38	1648.00 ± 1.48	230.00 ± 0.33	648.33 ± 0.16	54.08 ± 0.12	49.10 ± 0.13	1.40 ± 0.01	11.50 ± 0.28	6.80 ± 0.12
23.	*Dioscorea oppositifolia* var. *oppositifolia* (Tubers)	110.18 ± 0.14	1561.00 ± 0.98	880.60 ± 0.41	530.48 ± 0.12	88.46 ± 0.22	32.00 ± 0.51	5.24 ± 0.13	2.78 ± 0.01	8.40 ± 0.04
24.	*Dioscorea pentaphylla* var. *pentaphylla* (Tubers)	85.24 ± 0.11	1341.60 ± 1.41	640.10 ± 0.54	440.00 ± 0.32	126.10 ± 1.01	113.48 ± 0.12	3.22 ± 0.11	16.60 ± 0.13	2.32 ± 0.03
25.	*Dioscorea tomentosa* (Tubers)	35.00 ± 0.08	1345.41 ± 2.31	240.30 ± 0.13	192.00 ± 0.04	98.68 ± 0.62	23.66 ± 0.04	6.20 ± 0.12	1.44 ± 0.01	1.10 ± 0.01
26.	*Dioscorea spicata* (Tubers)	66.34 ± 0.54	1136.12 ± 0.74	234.10 ± 0.58	324.16 ± 0.24	166.30 ± 0.27	24.10 ± 0.26	2.56 ± 0.04	7.41 ± 0.11	6.70 ± 0.14
27.	*Dioscorea wallichi* (Tubers)	63.01 ± 0.27	1361.70 ± 1.01	748.31 ± 0.32	578.06 ± 0.19	106.40 ± 0.11	20.10 ± 0.04	6.66 ± 0.01	2.46 ± 0.08	3.31 ± 0.05
28.	*Dolichos trilobus* (Tubers)	1.48.33 ± 0.34	775.10 ± 1.21	680.00 ± 0.78	620.00 ± 0.11	115.10 ± 0.66	16.60 ± 0.23	4.44 ± 0.03	2.70 ± 0.06	22.14 ± 0.08
29.	*Hemidesmus indicus* var. *indicus* (Root)	26.10 ± 0.11	1065.04 ± 0.74	462.00 ± 0.31	84.21 ± 0.08	64.26 ± 0.10	42.18 ± 0.18	2.18 ± 0.02	1.12 ± 0.02	1.46 ± 0.01
30.	*Hemidesmus indicus* var. *pubescens* (Root)	41.02 ± 0.18	946.22 ± 0.66	438.41 ± 0.21	256.08 ± 0.24	106.20 ± 0.24	33.08 ± 0.10	1.44 ± 0.01	0.94 ± 0.03	1.84 ± 0.03
31.	*Ipomoea staphylina* (Root)	33.48 ± 0.68	620.30 ± 0.12	520.14 ± 0.18	410.10 ± 0.48	84.46 ± 0.34	31.20 ± 0.13	16.66 ± 0.12	1.42 ± 0.01	13.80 ± 0.03
32.	*Kedrostis foetidissima* (Tubers)	40.38 ± 0.68	718.41 ± 0.24	368.18 ± 0.22	154.18 ± 0.17	86.02 ± 0.16	32.68 ± 0.68	1.66 ± 0.03	1.64 ± 0.01	1.54 ± 0.03
33.	*Maerua oblongifolia* (Tubers)	66.00 ± 0.13	843.26 ± 0.32	218.16 ± 0.14	118.24 ± 0.13	76.06 ± 0.08	29.10 ± 0.06	1.44 ± 0.02	1.10 ± 0.02	1.32 ± 0.02
34.	*Manihot esculenta* (Tubers)	33.18 ± 0.10	940.00 ± 0.10	420.14 ± 0.07	310.10 ± 0.07	82.00 ± 0.03	28.20 ± 0.02	8.00 ± 0.07	1.20 ± 0.01	9.10 ± 0.03
35.	*Maranta arundinacea* (Rhizome)	22.40 ± 0.03	880.10 ± 0.68	440.70 ± 0.11	312.00 ± 0.110	54.00 ± 0.06	18.30 ± 0.05	3.60 ± 0.03	1.26 ± 0.01	7.54 ± 0.04
36.	*Momordica dioica* (Tubers)	85.00 ± 0.12	938.26 ± 0.24	234.08 ± 0.08	112.6 ± 0.31	64.08 ± 0.06	34.14 ± 0.31	1.24 ± 0.04	1.10 ± 0.01	1.58 ± 0.01

TABLE 5.14 *(Continued)*

Sl. No.	Botanical Name	Sodium	Potassium	Calcium	Magnesium	Phosphorus	Iron	Zinc	Copper	Manganese
37.	*Nephrolepis auriculata* (Tubers)	51.36 ± 0.14	1036.18 ± 0.12	236.54 ± 1.01	286.36 ± 0.32	69.32 ± 0.07	18.30 ± 0.54	2.01 ± 0.03	2.86 ± 0.03	4.36 ± 0.15
38.	*Nymphaea pubescens* (Tubers)	56.10 ± 0.22	768.14 ± 0.48	326.10 ± 0.17	96.18 ± 0.14	54.10 ± 0.11	32.10 ± 0.16	1.34 ± 0.01	1.16 ± 0.05	1.38 ± 0.02
	Nymphaea rubra (Tubers)	34.10 ± 0.16	734.00 ± 0.62	354.16 ± 0.32	104.08 ± 0.05	76.30 ± 0.04	28.14 ± 0.24	1.64 ± 0.01	1.12 ± 0.01	1.34 ± 0.01
	Parthenocissus neilgherriensis (Tubers)	48.09 ± 0.14	968.36 ± 0.65	372.14 ± 0.08	104.12 ± 0.15	72.16 ± 0.06	29.00 ± 0.13	1.04 ± 0.01	1.14 ± 0.01	1.64 ± 0.03
	Sarcostemma acidum (Tubers)	24.30 ± 0.18	946.00 ± 1.02	120.30 ± 0.54	138.72 ± 1.04	134.00 ± 0.72	52.00 ± 0.32	6.80 ± 0.24	1.40 ± 0.03	6.48 ± 0.33
	Sterculia urens (Root)	28.40 ± 0.12	525.10 ± 0.15	130.00 ± 0.17	250.55 ± 0.33	35.18 ± 0.04	44.15 ± 0.04	1.34 ±0.04	1.56 ± 0.05	4.40 ± 0.02
	Xanthosoma segittifolium (Corm)	168.10 ± 0.03	2016.00 ± 0.71	670.10 ± 0.07	186.48 ± 0.03	108.00 ± 0.07	72.00 ± 0.03	2.10 ± 0.01	5.94 ± 0.01	3.18 ± 0.01
	Xanthosoma violaceum (Corm)	194.00 ± 0.01	1948.10 ± 0.68	658.40 ± 0.12	177.10 ± 0.08	102.10 ± 0.02	84.70 ± 0.04	2.28 ± 0.03	6.60 ± 0.03	4.34 ± 0.03

[a]All the values are means of triplicate determinations expressed on dry weight basis.

± Denotes standard error.

TABLE 5.15 Mineral Composition of Edible Greens (mg 100^{-1})[a]

Sl. No.	Botanical Name	Sodium	Potassium	Calcium	Magnesium	Phosphorus	Iron	Zinc	Copper	Maganese
1.	*Acacia caesia*	74.36 ± 0.54	538.10 ± 0.60	38.12 ± 0.09	54.60 ± 0.11	141.04 ± 0.13	4.38 ± 0.01	0.98 ± 0.01	0.98 ± 0.01	1.34 ± 0.09
2.	*Acacia grahamii*	54.32 ± 0.32	668.08 ± 0.58	42.16 ± 0.12	66.32 ± 0.14	121.08 ± 0.14	5.16 ± 0.03	0.78 ± 0.01	0.54 ± 0.02	1.24 ± 0.06
3.	*Achyranthus aspera*	44.16 ± 0.18	558 .14 ± 0.38	33.16 ± 0.11	44.32 ± 0.09	98.34 ± 0.42	4.11 ± 0.04	0.58 ± 0.03	0.56 ± 0.03	1.09 ± 0.11
4.	*Achyranthes bidentata*	46.40 ± 0.31	738.46 ± 0.64	112.42 ± 0.11	38.10 ± 0.12	154.10 ± 0.24	4.12 ± 0.01	1.21 ± 0.04	0.48 ± 0.01	1.06 ± 0.03
5.	*Allmania nodiflora var. angustifolia*	56.66 ± 0.12	509.38 ± 0.64	48.32 ± 0.09	32.33 ± 0.16	143.08 ± 0.12	3.38 ± 0.09	0.66 ± 0.03	0.33 ± 0.03	1.13 ± 0.15
6.	*Allmania nodiflora var. procumbens.*	98.38 ± 0.46	512.16 ± 0.34	66.54 ± 0.14	31.04 ± 0.12	94.36 ± 0.32	3.94 ± 0.06	1.02 ± 0.06	0.58 ± 0.04	1.16 ± 0.13
7.	*Aloe vera*	112.34 ± 0.64	934.32 ± 1.31	131.04 ± 0.62	56.66 ± 0.22	101.33 ± 0.56	4.53 ± 0.07	0.66 ± 0.03	0.42 ± 0.01	0.98 ± 0.02
8.	*Alternanthera bettzickiana*	94.36 ± 0.41	10.32.16 ± 0.98	92.14 ± 0.32	63.14 ± 0.14	90.56 ± 0.42	7.26 ± 0.04	0.72 ± 0.02	0.38 ± 0.02	1.11 ± 0.03
9.	*Alternanthera sessilis*	136.32 ± 0.32	1014.98 ± 0.36	58.36 ± 0.31	45.33 ± 0.32	83.33 ± 0.33	9.32 ± 0.06	0.98 ± 0.01	0.56 ± 0.04	1.18 ± 0.06
10.	*Amaranthus roxburghianus*	92.01 ± 0.11	1236.14 ± 1.01	66.32 ± 0.56	56.06 ± 0.14	121.14 ± 0.56	11.56 ± 0.08	0.86 ± 0.02	0.66 ± 0.01	1.18 ± 0.06
11.	*Amaranthus spinosus*	114.10 ± 0.21	1548.36 ± 0.79	317.12 ± 0.67	140.10 ± 0.16	90.10 ± 0.27	22.50 ± 0.11	1.14 ± 0.03	1.18 ± 0.03	2.30 ± 0.03
12.	*Amaranthus tricolor*	101.08 ± 0.32	1636.22 ± 0.84	184.16 ± 0.52	88.36 ± 0.18	101.08 ± 0.62	17.32 ± 0.08	1.01 ± 0.02	0.72 ± 0.01	1.76 ± 0.02
13.	*Amaranthus viridis.*	138.16 ± 0.33	1436.10 ± 0.32	2.24.32 ± 0.32	94.33 ± 0.16	98.98 ± 0.33	15.32 ± 0.09	0.96 ± 0.02	0.62 ± 0.03	1.66 ± 0.12
14.	*Amorphophallus sylvaticus*	66.34 ± 0.16	668.36 ± 0.78	98.11 ± 0.12	53.12 ± 0.11	68.32 ± 0.23	7.32 ± 0.11	0.94 ± 0.03	0.72 ± 0.05	1.24 ± 0.08
15.	*Asystasia gangetica*	47.10 ± 0.14	1038.48 ± 1.32	196.32 ± 0.04	112.14 ± 0.31	96.40 ± 0.09	6.36 ± 0.03	0.78 ± 0.01	1.12 ± 0.01	2.12 ± 0.01
16.	*Basella alba*	141.38 ± 0.09	2434.10 ± 1.71	224.34 ± 0.03	78.32 ± 0.17	199.30 ± 0.10	21.30 ± 0.07	1.48 ± 0.04	1.10 ± 0.01	1.36 ± 0.01
17.	*Begonia malabarica*	65.20 ± 0.14	986.36 ± 0.98	96.12 ± 0.04	63.12 ± 0.11	154.15 ± 0.08	5.46 ± 0.04	0.98 ± 0.02	0.66 ± 0.02	1.11 ± 0.02
18.	*Boerahavia diffusa*	56.15 ± 0.08	666.12 ± 0.54	88.06 ± 0.07	56.66 ± 0.08	166.14 ± 0.06	4.12 ± 0.06	1.04 ± 0.03	0.88 ± 0.03	0.94 ± 0.01
19.	*Boerhavia erecta*	72.33 ± 0.17	534.18 ± 0.18	111.36 ± 0.09	33.08 ± 0.12	182.12 ± 0.11	5.21 ± 0.03	0.86 ± 0.02	0.72 ± 0.01	1.04 ± 0.01
20.	*Borassus flabellifer*	38.08 ± 0.09	1038.24 ± 0.32	124.06 ± 0.10	96.12 ± 0.16	176.62 ± 0.15	11.24 ± 0.06	0.76 ± 0.03	0.58 ± 0.07	0.96 ± 0.08

TABLE 5.15 *(Continued)*

Sl. No.	Botanical Name	Sodium	Potassium	Calcium	Magnesium	Phosphorus	Iron	Zinc	Copper	Maganese
21.	*Brassica juncea*	34.40 ± 0.13	1948.12 ± 1.42	97.20 ± 0.06	124.10 ± 0.15	190.48 ± 0.42	9.39 ± 0.12	0.48 ± 0.01	0.92 ± 0.01	1.21 ± 0.01
22.	*Canthium parvifolium*	46.14 ± 0.12	856.13 ± 0.56	78.26 ± 0.09	92.04 ± 0.14	112.09 ± 0.21	6.21 ± 0.07	0.38 ± 0.01	0.72 ± 0.01	0.68 ± 0.02
23.	*Cassia obtusifolia*	62.06 ± 0.32	568.33 ± 0.33	81.03 ± 0.06	86.12 ± 0.09	132.16 ± 0.36	4.11 ± 0.04	0.42 ± 0.02	0.88 ± 0.04	0.74 ± 0.03
24.	*Cassia tora*	34.12 ± 0.10	538.36 ± 0.78	33.26 ± 0.12	44.23 ± 0.16	112.14 ± 0.21	6.14 ± 0.03	0.78 ± 0.01	0.66 ± 0.01	0.98 ± 0.06
25.	*Capsium annum*	31.64 ± 0.12	632.60 ± 0.31	78.00 ± 0.13	38.00 ± 0.13	117.16 ± 0.28	5.08 ± 0.07	2.74 ± 0.03	1.18 ± 0.01	1.36 ± 0.06
26.	*Capsicum frutescens*	42.36 ± 0.08	612.30 ± 0.21	46.06 ± 0.08	52.06 ± 0.08	101.01 ± 0.18	5.12 ± 0.02	1.01 ± 0.02	0.78 ± 0.01	1.12 ± 0.03
27.	*Cardiospermum canescens*	56.14 ± 0.12	736.18 ± 0.32	52.11 ± 0.11	46.11 ± 0.09	114.02 ± 0.24	8.33 ± 0.12	0.98 ± 0.01	0.56 ± 0.03	1.16 ± 0.04
28.	*Cardiospermum helicacabum*	52.08 ± 0.09	886.20 ± 0.14	54.36 ± 0.06	32.12 ± 0.11	98.26 ± 0.26	9.12 ± 0.08	0.86 ± 0.02	0.68 ± 0.02	0.86 ± 0.02
29.	*Cardiospermum microcarpa*	66.26 ± 0.11	786.14 ± 0.20	33.10 ± 0.08	33.26 ± 0.09	101.21 ± 0.14	5.36 ± 0.03	0.33 ± 0.01	0.88 ± 0.01	1.11 ± 0.02
30.	*Celosia argentea*	32.06 ± 0.14	824.26 ± 0.31	44.12 ± 0.11	42.06 ± 0.14	115.17 ± 0.17	5.42 ± 0.05	0.92 ± 0.02	0.92 ± 0.03	0.96 ± 0.01
31.	*Centella asiatica*	84.14 ± 0.12	936.02 ± 0.28	56.46 ± 0.09	53.13 ± 0.17	124.06 ± 0.22	6.66 ± 0.15	1.11 ± 0.03	0.86 ± 0.03	1.22 ± 0.03
32.	*Cissus quadrangularis*	50.12 ± 0.08	852.14 ± 0.32	66.32 ± 0.14	46.33 ± 0.13	131.16 ± 0.33	5.11 ± 0.16	1.74 ± 0.06	0.72 ± 0.01	1.21 ± 0.05
33.	*Cissus vitiginea*	41.06 ± 0.12	854.03 ± 0.44	111.32 ± 0.08	66.16 ± 0.05	101.54 ± 0.08	5.36 ± 0.03	0.98 ± 0.01	0.94 ± 0.01	0.93 ± 0.02
34.	*Cleome gynandra*	30.11± 0.06	866.05 ± 0.32	96.33 ± 0.11	44.32 ± 0.03	86.52 ± 0.06	5.12 ± 0.02	1.01 ± 0.02	0.87 ± 0.02	0.96 ± 0.01
35.	*Cleome viscosa*	28.13 ± 0.03	534.06 ± 0.18	101.41 ± 0.09	53.12 ± 0.04	92.12 ± 0.04	6.24 ± 0.01	0.89 ± 0.03	0.92 ± 0.01	1.04 ± 0.03
36.	*Coccinia grandis*	54.26 ± 0.02	638.12 ± 0.24	88.52 ± 0.16	58.52 ± 0.06	102.16 ± 0.09	6.08 ± 0.03	1.11 ± 0.01	0.86 ± 0.03	1.16 ± 0.05
37.	*Cocculus hirsutus*	66.06 ± 0.11	768.33 ± 0.11	92.16 ± 0.14	66.72 ± 0.08	106.32 ± 0.08	7.34 ± 0.04	0.82 ± 0.03	0.82 ± 0.02	1.08 ± 0.04
38.	*Colocasia esculenta*	58.24 ± 0.06	936.32 ± 0.33	94.33 ± 0.52	74.12 ± 0.09	118.20 ± 0.14	11.14 ± 0.03	0.96 ± 0.01	0.72 ± 0.03	1.11 ± 0.02
39.	*Commelina benghalensis*	86.48 ± 0.11	942.10 ± 0.51	112.40 ± 0.12	87.36 ± 0.11	132.00 ± 0.11	9.32 ± 0.01	1.12 ± 0.01	0.68 ± 0.01	0.98 ± 0.01
40.	*Commelina ensifolia*	72.12 ± 0.08	814.08 ± 0.33	121.16 ± 0.08	85.16 ± 0.03	124.32 ± 0.08	8.54 ± 0.03	0.99 ± 0.05	0.66 ± 0.02	0.92 ± 0.03
41.	*Coriandrum sativum*	53.16 ± 0.06	916.02 ± 0.18	88.12 ± 0.11	56.34 ± 0.06	114.12 ± 0.06	6.31 ± 0.02	0.76 ± 0.01	0.56 ± 0.01	1.01 ± 0.02

TABLE 5.15 *(Continued)*

Sl. No.	Botanical Name	Sodium	Potassium	Calcium	Magnesium	Phosphorus	Iron	Zinc	Copper	Maganese
42.	*Cycas circinalis*	41.26 ± 0.06	632.14 ± 1.16	115.17 ± 0.13	66.32 ± 0.21	98.34 ± 0.32	9.18 ± 0.08	1.11 ± 0.03	0.88 ± 0.01	0.94 ± 0.02
43.	*Digera muricata*	33.06 ± 0.04	714.10 ± 0.71	116.32 ± 0.16	52.16 ± 0.14	132.06 ± 0.24	10.36 ± 0.11	0.96 ± 0.02	1.21 ± 0.02	1.16 ± 0.03
44.	*Diplocyclos palmatus*	34.16 ± 0.09	946.32 ± 0.74	132.40 ± 0.31	47.10 ± 0.31	170.40 ± 0.29	17.14 ± 0.14	0.92 ± 0.06	0.94 ± 0.03	1.32 ± 0.01
45.	*Emilia sonchifolia*	50.16 ± 0.07	886.12 ± 0.54	111.02 ± 0.21	49.26 ± 0.17	121.14 ± 0.06	5.26 ± 0.03	0.72 ± 0.03	0.98 ± 0.02	1.20 ± 0.03
46.	*Euphorbia hirta*	41.11 ± 0.06	766.11 ± 0.33	98.03 ± 0.11	55.24 ± 0.24	111.26 ± 0.24	5.36 ± 0.03	0.91 ± 0.01	0.88 ± 0.01	1.08 ± 0.02
47.	*Gisekia pharnaceoides*	38.26 ± 0.05	754.14 ± 0.22	117.16 ± 0.08	66.18 ± 0.07	86.21 ± 0.07	6.72 ± 0.05	0.88 ± 0.02	1.16 ± 0.04	1.11 ± 0.06
48.	*Glinus oppositifolius*	43.15 ± 0.03	684.26 ± 0.10	76.36 ± 0.04	54.11 ± 0.06	77.06 ± 0.14	7.76 ± 0.06	1.01 ± 0.03	0.92 ± 0.03	1.21 ± 0.04
49.	*Heracleum rigens var. rigens*	52.16 ± 0.02	732.16 ± 0.08	89.21 ± 0.06	33.24 ± 0.05	79.23 ± 0.11	7.02 ± 0.14	0.96 ± 0.02	0.96 ± 0.02	0.98 ± 0.05
50.	*Hybanthus enneaspermus*	41.08 ± 0.07	632.08 ± 0.12	46.12 ± 0.11	54.36 ± 0.12	112.32 ± 0.26	6.66 ± 0.09	0.92 ± 0.01	0.78 ± 0.02	0.88 ± 0.01
51.	*Ipomoea aquatica*	33.16 ± 0.01	672.52 ± 0.08	52.36 ± 0.09	66.38 ± 0.14	96.18 ± 0.18	7.33 ± 0.11	1.02 ± 0.02	0.66 ± 0.01	1.21 ± 0.04
52.	*Ipomoea pestigridis*	38.52 ± 0.06	732.14 ± 0.14	34.56 ± 0.04	58.32 ± 0.08	88.56 ± 0.12	5.23 ± 0.03	0.94 ± 0.03	0.72 ± 0.01	1.36 ± 0.06
53.	*Jasminum auriculatum*	52.34 ± 0.11	744.36 ± 0.21	38.78 ± 0.06	62.12 ± 0.09	76.33 ± 0.08	5.06 ± 0.04	0.88 ± 0.01	0.69 ± 0.02	1.26 ± 0.08
54.	*Jasminum calophyllum*	47.36 ± 0.09	982.08 ± 0.11	44.36 ± 0.08	46.32 ± 0.11	92.36 ± 0.08	4.44 ± 0.10	0.96 ± 0.01	0.82 ± 0.03	1.11 ± 0.06
55.	*Kalanchoe pinnata*	56.66 ± 0.08	638.12 ± 0.08	68.32 ± 0.07	78.01 ± 0.04	106.42 ± 0.17	6.86 ± 0.11	0.84 ± 0.02	0.92 ± 0.04	0.86 ± 0.03
56.	*Lathyrus sativus*	66.32 ± 0.11	786.34 ± 0.32	132.16 ± 0.08	82.56 ± 0.06	74.12 ± 0.12	5.10 ± 0.06	0.91 ± 0.02	0.94 ± 0.01	1.64 ± 0.05
57.	*Leucas montana var. wightii.*	18.32 ± 0.04	532.16 ± 0.51	116.34 ± 0.21	69.30 ± 0.38	98.16 ± 0.08	6.64 ± 0.11	0.74 ± 0.03	0.67 ± 0.01	2.12 ± 0.03
58.	*Mollugo pentaphylla*	24.32 ± 0.06	589.09 ± 0.36	89.26 ± 0.12	76.34 ± 0.24	82.66 ± 0.15	5.26 ± 0.08	0.92 ± 0.03	0.78 ± 0.02	1.04 ± 0.02
59.	*Moringa concanensis*	82.08 ± 0.26	1568.08 ± 0.66	121.12 ± 0.14	54.32 ± 0.11	128.06 ± 0.32	15.2 ± 0.08	0.88 ± 0.02	0.92 ± 0.01	1.02 ± 0.02
60.	*Mukia maderaspatana*	112.46 ± 0.54	1738.10 ± 0.57	142.42 ± 0.11	78.40 ± 0.21	189.00 ± 0.11	17.14 ± 0.14	0.48 ± 0.01	0.94 ± 0.03	1.32 ± 0.01
61.	*Murraya koenigii*	66.12 ± 0.11	1416.08 ± 0.31	114.08 ± 0.08	44.06 ± 0.08	132.16 ± 0.09	16.66 ± 0.11	0.66 ± 0.03	0.88 ± 0.01	1.16 ± 0.04
62.	*Murraya paniculata*	52.58 ± 0.14	1366.11 ± 0.24	108.21 ± 0.06	52.16 ± 0.06	126.32 ± 0.14	14.30 ± 0.09	0.76 ± 0.02	0.79 ± 0.02	1.52 ± 0.06

TABLE 5.15 *(Continued)*

Sl. No.	Botanical Name	Sodium	Potassium	Calcium	Magnesium	Phosphorus	Iron	Zinc	Copper	Maganese
63.	*Oxalis corniculata*	46.30 ± 0.23	1624.12 ± 0.77	112.72 ± 0.2	54.40 ± 0.17	109.58 ± 0.21	9.68 ± 0.14	1.68 ± 0.01	0.74 ± 0.01	1.74 ± 0.01
64.	*Oxalis latifolia*	66.36 ± 0.21	982.16 ± 0.24	98.14 ± 0.12	66.38 ± 0.11	98.26 ± 0.09	8.88 ± 0.08	0.94 ± 0.06	0.91 ± 0.03	0.98 ± 0.03
65.	*Peperomia pellucida*	72.04 ± 0.14	832.12 ± 0.12	106.11 ± 0.16	98.26 ± 0.06	101.11 ± 0.06	7.06 ± 0.06	0.56 ± 0.02	0.73 ± 0.02	0.88 ± 0.03
66.	*Physalis minima var.indica*	76.52 ± 0.11	1022.10 ± 0.08	96.04 ± 0.12	111.26 ± 0.08	124.16 ± 0.18	8.12 ± 0.04	0.92 ± 0.01	0.84 ± 0.01	0.72 ± 0.02
67.	*Portulaca oleracea var oleracea*	40.32 ± 0.08	854.32 ± 0.11	68.36 ± 0.08	78.32 ± 0.11	98.36 ± 0.14	5.62 ± 0.02	0.96 ± 0.02	0.94 ± 0.01	0.96 ± 0.01
68.	*Portulaca quadrifida*	55.16 ± 0.12	862.10 ± 0.21	56.06 ± 0.06	82.14 ± 0.16	120.08 ± 0.17	5.98 ± 0.01	0.88 ± 0.01	0.91 ± 0.03	10.8 ± 0.03
69.	*Premna corymbosa*	32.18 ± 0.06	756.52 ± 0.08	74.11 ± 0.09	56.66 ± 0.14	118.16 ± 0.11	6.06 ± 0.03	0.92 ± 0.02	0.84 ± 0.02	0.94 ± 0.02
70.	*Psilanthus wightianus*	41.23 ± 0.04	766.14 ± 0.06	66.30 ± 0.04	70.54 ± 0.12	58.36 ± 0.08	7.14 ± 0.02	0.94 ± 0.01	0.76 ± 0.04	1.14 ± 0.05
71.	*Sarcostemna acidum*	52.06 ± 0.06	894.15 ± 0.06	114.16 ± 0.36	66.36 ± 0.08	78.14 ± 0.16	8.62 ± 0.04	1.01 ± 0.04	0.89 ± 0.03	1.06 ± 0.02
72.	*Sesbania grandiflora*	61.32 ± 0.08	1288.32 ± 0.28	128.08 ± 0.26	68.12 ± 0.06	112.32 ± 0.15	12.34 ± 0.11	0.76 ± 0.02	0.86 ± 0.02	1.26 ± 0.03
73.	*Solanum anguivi var. multiflora*	58.46 ± 0.11	1168.18 ± 0.32	116.14 ± 0.16	98.06 ± 0.17	108.11 ± 0.12	9.38 ± 0.08	0.82 ± 0.01	0.99 ± 0.04	4.20 ± 0.04
74.	*Solanum nigrum*	108.32 ± 0.27	1456.36 ± 0.73	47.10 ± 0.14	87.12 ± 0.11	186.10 ± 0.31	10.30 ± 0.09	1.12 ± 0.01	1.10 ± 0.04	7.42 ± 0.17
75.	*Solanum trilobatum*	132.16 ± 0.21	1624.14 ± 0.88	54.08 ± 0.11	99.36 ± 0.10	155.08 ± 0.21	9.36 ± 0.08	0.96 ± 0.02	0.98 ± 0.02	5.26 ± 0.14
76.	*Tamarindus indica*	114.06 ± 0.15	1431.26 ± 0.66	60.56 ± 0.08	110.12 ± 0.08	166.54 ± 0.36	12.32 ± 0.04	1.04 ± 0.03	0.76 ± 0.01	2.15 ± 0.11
77.	*Timospora cordifolia*	94.52 ± 0.18	982.51 ± 0.32	49.30 ± 0.10	94.16 ± 0.06	98.02 ± 0.18	9.32 ± 0.03	0.94 ± 0.01	0.88 ± 0.03	0.96 ± 0.04
78.	*Trianthema portulacastrum*	126.21 ± 0.32	1108.50 ± 0.41	50.16 ± 0.08	92.08 ± 0.04	108.26 ± 0.19	8.28 ± 0.03	0.66 ± 0.02	0.92 ± 0.01	0.78 ± 0.03
79.	*Vigna radiate var. sublobata*	111.09 ± 0.18	1126.26 ± 0.34	48.16 ± 0.04	87.72 ± 0.06	121.11 ± 0.24	6.24 ± 0.02	0.88 ± 0.01	1.03 ± 0.03	0.99 ± 0.02
80.	*Vigna trilobata*	103.14 ± 0.16	1084.14 ± 0.16	44.32 ± 0.06	91.06 ± 0.04	132.16 ± 0.34	5.98 ± 0.03	0.92 ± 0.02	0.95 ± 0.01	1.26 ± 0.05

[a] All the values are means of triplicate determinations expressed on dry weight basis.
± Denotes standard error.

TABLE 5.16 Mineral Composition of Edible Seeds and Seed Components (mg 100^{-1})[a]

Sl. No.	Botanical Name	Sodium	Potassium	Calcium	Magnesium	Phosphorus	Iron	Zinc	Copper	Maganese
1.	*Artocarpus heterophyllus* (Kernel)	32.06 ± 0.22	1532.01 ± 0.96	112.06 ± 0.13	98.34 ± 0.22	113.26 ± 0.18	21.06 ± 0.06	1.48 ± 0.03	0.98 ± 0.01	1.01 ± 0.02
2.	*Atylosia scarabaeoides* (Seed)	28.50 ± 0.54	1884.32 ± 1.08	164.24 ± 0.11	126.30 ± 0.43	168.14 ± 0.12	42.14 ± 0.14	2.70 ± 0.01	1.32 ± 0.12	0.96 ± 0.01
3.	*Bambusa arundinacea* (Seed)	44.22 ± 0.18	1032.56 ± 0.78	168.36 ± 0.28	144.36 ± 0.32	98.22 ± 0.08	32.33 ± 0.06	2.84 ± 0.06	1.14 ± 0.08	1.26 ± 0.03
4.	*Borassus flabellifer* (Endosperm)	33.04 ± 0.21	1396.18 ± 0.32	212.08 ± 0.21	108.16 ± 0.18	124.06 ± 0.09	30.16 ± 0.11	4.44 ± 0.06	2.56 ± 0.04	2.24 ± 0.04
5.	*Bupleurum wightii*.var. *ramosissimum* (Seed)	24.36 ± 0.11	838.06 ± 0.94	236.15 ± 0.32	124.56 ± 0.36	136.05 ± 0.15	28.14 ± 0.08	1.28 ± 0.03	0.76 ± 0.05	2.86 ± 0.11
6.	*Canarium strictum* (Kernel)	22.54 ± 0.08	950.36 ± 0.12	158.00 ± 0.04	132.00 ± 0.11	88.30 ± 0.94	24.70 ± 0.07	1.30 ± 0.01	2.10 ± 0.01	3.80 ± 0.08
7.	*Canavalia gladiata* (Seed)	28.06 ± 0.50	2216.31 ± 1.30	280.00 ± 0.58	74.06 ± 0.18	266.30 ± 0.19	45.20 ± 0.11	3.70 ± 0.03	0.98 ± 0.01	0.84 ± 0.01
8.	*Canavalia virosa* (Seed)	34.14 ± 0.24	1938.08 ± 1.16	254.30 ± 0.36	112.36 ± 0.19	318.06 ± 0.20	38.14 ± 0.08	4.12 ± 0.05	1.21 ± 0.01	1.33 ± 0.04
9.	*Capparis zeylanica* (Seed)	21.08 ± 0.03	786.30 ± 0.26	112.16 ± 0.03	232.16 ± 2.46	136.56 ± 0.28	3.30 ± 0.01	1.26 ± 0.01	1.12 ± 0.02	3.21 ± 0.04
10.	*Celtis philippensis* var. wightii (Seed)	14.62 ± 0.04	666.14 ± 0.14	214.06 ± 0.08	148.36 ± 0.78	168.24 ± 0.24	4.44 ± 0.03	1.56 ± 0.03	1.08 ± 0.03	2.48 ± 0.03
11.	*Chamaecrista absus* (Seed)	24.16 ± 0.08	324.24 ± 0.17	248.36 ± 0.94	218.46 ± 0.94	224.06 ± 0.31	9.32 ± 0.04	2.28 ± 0.06	0.96 ± 0.01	1.22 ± 0.04
12.	*Cycas circinalis* (Kernel)	27.14 ± 0.012	984.50 ± 0.31	234.00 ± 0.08	240.10 ± 0.28	124.10 ± 0.31	21.30 ± 0.12	3.40 ± 0.02	1.20 ± 0.01	1.10 ± 0.01
13.	*Dolichos lablab* var.*Vulgaris* (seed)	24.05 ± 0.14	1724.08 ± 0.24	224.16 ± 0.07	78.48 ± 0.06	168.42 ± 0.22	15.24 ± 0.14	1.98 ± 0.02	058 ± 0.01	1.38 ± 0.01
14.	*Dolichos trilobus* (Seed)	26.50 ± 0.52	1856.10 ± 0.36	260.50 ± 0.33	66.30 ± 0.07	154.10 ± 0.28	18.36 ± 0.18	2.10 ± 0.01	0.48 ± 0.01	1.26 ± 0.01
15.	*Drypetes sepiaria* (Seed)	31.24 ± 0.62	1126.20 ± 1.21	246.28 ± 0.18	212.08 ± 0.03	164.30 ± 0.18	16.26 ± 0.08	2.24 ± 0.08	0.33 ± 0.02	0.98 ± 0.26
16.	*Elaeocarpus tectorius* (Kernel)	18.24 ± 0.04	680.00 ± 0.82	280.00 ± 0.28	138.14 ± 0.16	128.18 ± 0.33	17.80 ± 0.12	4.10 ± 0.01	2.40 ± 0.01	5.30 ± 0.01
17.	*Eleusine coracana* (Seed)	25.56 ± 0.14	798.42 ± 0.14	296.76 ± 0.14	154.36 ± 0.22	144.36 ± 0.21	15.32 ± 0.14	3.36 ± 0.05	1.84 ± 0.04	4.46 ± 0.04
18.	*Ensete superbum* (Seed)	33.20 ± 0.17	668.00 ± 0.76	233.52 ± 14	332.08 ± 0.18	130.10 ± 0.14	18.16 ± 0.04	3.58 ± 0.07	1.26 ± 0.03	3.32 ± 0.02
19.	*Entada rheedi* (Kernel)	35.00 ± 0.14	2637.72 ± 1.50	284.00 ± 0.31	640.26 ± 0.08	244.10 ± 0.11	22.26 ± 0.12	1.12 ± 0.01	1.24 ± 0.04	5.24 ± 0.05
20.	*Givotia rottleriformis* (Kernel)	35.18 ± 0.32	1035.18 ± 0.72	408.38 ± 0.11	168.31 ± 0.34	138.40 ± 0.36	29.16 ± 0.11	2.30 ± 0.04	1.90 ± 0.01	6.50 ± 0.03
21.	*Heracleum rigens* var. *rigens* (Seed)	18.24 ± 0.03	728.16 ± 0.22	212.04 ± 0.28	168.24 ± 0.14	256.26 ± 0.18	8.11 ± 0.11	1.12 ± 0.06	1.01 ± 0.01	2.24 ± 0.03

TABLE 5.16 *(Continued)*

Sl. No.	Botanical Name	Sodium	Potassium	Calcium	Magnesium	Phosphorus	Iron	Zinc	Copper	Maganese
22.	*Impatiens balsamina* (Seed)	28.26 ± 0.14	948.32 ± 0.36	224.30 ± 0.30	228.08 ± 0.08	276.14 ± 0.09	4.22 ± 0.03	0.98 ± 0.03	1.24 ± 0.06	1.98 ± 0.26
23.	*Lablab purpureus* var. *lignosus* (Seed)	55.15 ± 0.24	2438.54 ± 1.78	520.00 ± 0.38	620.00 ± 0.18	268.12 ± 0.32	7.90 ± 0.33	2.14 ± 0.06	1.58 ± 0.02	6.54 ± 0.13
24.	*Lablab purpureus* var. *purpureus* (Seed)	66.08 ± 0.41	1936.16 ± 1.28	498.15 ± 0.18	514.18 ± 0.14	272.10 ± 0.28	14.32 ± 0.18	1.08 ± 0.04	1.66 ± 0.03	5.21 ± 0.14
25.	*Macrotyloma uniflorum* (seed)	18.16 ± 0.22	1236.14 ± 0.34	154.36 ± 0.16	204.06 ± 0.24	114.36 ± 0.11	7.26 ± 0.05	2.68 ± 0.02	0.68 ± 0.01	1.78 ± 0.02
26.	*Moringa concanensis* (Kernel)	37.50 ± 0.34	1155.00 ± 0.08	320.00 ± 0.12	244.34 ± 0.31	154.30 ± 0.78	34.50 ± 0.14	3.10 ± 0.01	1.64 ± 0.01	0.60 ± 0.01
27.	*Mucuna atropurpurea* (Kernel)	20.00 ± 0.14	2139.34 ± 1.33	189.69 ± 0.34	98.88 ± 0.18	152.18 ± 0.12	5.65 ± 0.08	3.13 ± 0.05	0.88 ± 0.01	5.96 ± 0.04
28.	*Mucuna pruriens* var. *pruriens* (Seed)	66.64 ± 0.030	1.662.12 ± 0.52	738.16 ± 0.41	514.16 ± 2.06	418.08 ± 0.13	5.47 ± 0.04	2.24 ± 0.08	0.44 ± 0.03	7.14 ± 0.03
29.	*Mucuna pruriens* var. *utilis* (Black coloured seed coat) (Seed)	86.44 ± 0.03	1933.14 ± 0.63	723.16 ± 0.41	268.30 ± 0.21	310.18 ± 0.14	11.08 ± 0.03	7.15 ± 0.02	1.86 ± 0.04	3.67 ± 0.02
30.	*Mucuna pruriens* var. *utilis* (White coloured seed coat)	114.08 ± 0.11	1728.36 ± 0.06	748.30 ± 0.19	352.64 ± 0.08	466.32 ± 0.17	12.72 ± 0.06	4.11 ± 0.06	2.01 ± 0.03	4.21 ± 0.02
31.	*Neonotonia wightii*. var. *coimbatorensis*. (Seed)	43.00 ± 0.12	1636.18 ± 1.24	280.20 ± 1.03	228.00 ± 0.83	184.15 ± 0.35	16.60 ± 0.12	2.40 ± 0.01	1.32 ± 0.03	1.60 ± 0.02
32.	*Ocimum gratissimum* (Seed)	15.35 ± 0.11	880.00 ± 0.44	188.30 ± 0.04	436.34 ± 0.08	84.50 ± 0.08	30.50 ± 0.13	2.10 ± 0.01	5.04 ± 0.02	1.74 ± 0.02
33.	*Oryza meyeriana* var. *granulata* (Grain)	21.31 ± 0.08	936.14 ± 0.21	220.18 ± 0.06	214.22 ± 0.18	136.24 ± 0.21	18.26 ± 0.08	1.24 ± 0.03	2.12 ± 0.03	1.86 ± 0.032
34.	*Pithecellobium dulce* (Aril)	14.14 ± 0.06	786.24 ± 0.26	184.20 ± 0.03	194.18 ± 0.12	144.14 ± 0.14	4.26 ± 0.21	2.12 ± 0.08	0.94 ± 0.02	2.33 ± 0.14
35.	*Rhynchosia cana* (Seed)	21.50 ± 0.08	1562.20 ± 0.88	160.30 ± 0.78	160.54 ± 0.75	174.10 ± 0.32	6.24 ± 0.14	4.14 ± 0.11	1.80 ± 0.01	8.40 ± 0.13
36.	*Rhynchosia filipes* (Seed)	25.00 ± 0.78	2128.18 ± 1.52	130.10 ± 0.08	134.15 ± 0.11	144.28 ± 0.08	33.10 ± 0.32	4.12 ± 0.01	1.50 ± 0.01	1.24 ± 0.02
37.	*Rhynchosia rufescens* (Seed)	34.14 ± 0.13	1833.14 ± 0.42	193.08 ± 0.28	168.42 ± 0.26	248.00 ± 0.17	9.41 ± 0.02	4.68 ± 0.01	1.68 ± 0.02	4.66 ± 0.03
38.	*Rhynchosia suaveolens* (Seed)	24.30 ± 0.76	1678.20 ± 0.94	210.24 ± 0.30	94.34 ± 0.60	278.15 ± 0.21	5.30 ± 0.16	3.54 ± 0.01	1.34 ± 0.10	7.36 ± 0.11

TABLE 5.16 *(Continued)*

Sl. No.	Botanical Name	Sodium	Potassium	Calcium	Magnesium	Phosphorus	Iron	Zinc	Copper	Maganese
39.	*Sapindus emarginatus* (Seed)	16.36 ± 0.03	786.16 ± 0.11	148.28 ± 0.08	248.14 ± 0.06	166.24 ± 0.09	2.48 ± 0.03	0.98 ± 0.01	1.28 ± 0.03	1.56 ± 0.07
40.	*Sesamum indicum* (Seed)	10.10 ± 0.05	687.34 ± 0.21	280.10 ± 0.03	298.31 ± 0.11	152.30 ± 0.18	1.40 ± 0.01	1.90 ± 0.01	2.10 ± 0.02	1.60 ± 0.08
41.	*Shorea roxburghii* (Cotyledon)	9.38 ± 0.06	946.12 ± 0.18	148.24 ± 0.04	224.08 ± 0.08	144.08 ± 0.09	7.28 ± 0.06	1.20 ± 0.02	0.98 ± 0.04	2.22 ± 0.08
42.	*Sterculia foetida* (Kernel)	35.34 ± 0.14	1104.00 ± 0.84	320.10 ± 0.12	544.00 ± 0.19	168.30 ± 0.14	16.60 ± 0.03	2.34 ± 0.02	1.80 ± 0.03	2.66 ± 0.12
43.	*Sterculia guttata* (Kernel)	17.55 ± 0.08	948.10 ± 0.34	348.08 ± 0.11	120.00 ± 0.02	120.10 ± 0.10	19.84 ± 0.11	1.30 ± 0.01	2.40 ± 0.40	0.80 ± 0.01
44.	*Sterculia urens* (Kernel)	22.50 ± 0.13	836.30 ± 0.31	240.00 ± 0.04	144.00 ± 0.03	134.00 ± 0.07	31.30 ± 0.13	3.60 ± 0.02	2.80 ± 0.01	0.70 ± 0.01
45.	*Strychnos nux-vomica* (Kernel)	23.00 ± 0.14	550.48 ± 0.54	375.10 ± 0.31	380.10 ± 0.18	148.00 ± 0.08	38.00 ± 0.16	4.61 ± 0.12	1.94 ± 0.01	4.40 ± 0.07
46.	*Tamarindus indica* (Kernel)	25.60 ± 0.34	20.38.10 ± 1.34	240.00 ± 0.19	248.00 ± 0.68	153.18 ± 0.16	9.20 ± 0.07	4.20 ± 0.03	2.40 ± 0.11	3.36 ± 0.03
47.	*Teramnus labialis* (Seed)	23.00 ± 0.22	1866.18 ± 1.06	230.20 ± 0.18	520.30 ± 0.86	162.10 ± 0.12	44.00 ± 0.12	3.20 ± 0.03	2.20 ± 0.08	0.48 ± 0.01
48.	*Terminalia bellirica* (Kernel)	22.24 ± 0.11	752.00 ± 0.14	410.30 ± 0.01	152.33 ± 0.38	124.10 ± 0.14	45.08 ± 0.18	2.60 ± 0.01	2.50 ± 0.01	1.50 ± 0.01
49.	*Terminalia chebula* (Kernel)	28.14 ± 0.06	883.10 ± 0.11	434.36 ± 0.07	166.12 ± 0.21	138.41 ± 0.13	24.36 ± 0.09	3.34 ± 0.03	1.84 ± 0.03	1.62 ± 0.01
50.	*Vigna bourneae* (Seed)	32.16 ± 0.13	2234.18 ± 1.36	284.26 ± 0.44	166.34 ± 0.08	173.52 ± 0.12	7.14 ± 0.05	0.98 ± 0.01	0.64 ± 0.03	0.84 ± 0.01
51.	*Vigna radiata* (Seed)	28.70 ± 0.34	2416.20 ± 0.88	244.36 ± 0.58	174.16 ± 0.98	146.38 ± 0.14	9.38 ± 0.03	1.12 ± 0.01	0.34 ± 0.02	0.58 ± 0.02
52.	*Vigna trilobata* (Seed)	35.56 ± 0.33	1964.20 ± 1.24	222.36 ± 0.37	152.00 ± 1.02	158.36 ± 0.12	0.42 ± 0.18	1.56 ± 0.01	0.88 ± 0.01	1.24 ± 0.03
53.	*Vigna unguiculata* subsp. *cylindrical* (Seed)	20.00 ± 0.10	1734.50 ± 0.74	240.52 ± 0.18	84.10 ± 0.34	166.30 ± 0.13	15.10 ± 0.01	3.90 ± 0.12	0.32 ± 0.01	1.50 ± 0.01
54.	*Vigna unguiculata* subsp. *unguiculata* (Seed)	32.50 ± 0.28	2018.92 ± 0.56	168.10 ± 0.18	24.00 ± 0.10	132.10 ± 0.06	14.92 ± 0.08	1.20 ± 0.01	0.64 ± 0.02	2.40 ± 0.04
55.	*Zizyphus rugosa* (Kernel)	21.38 ± 0.06	468.32 ± 0.12	268.08 ± 0.11	212.16 ± 0.14	98.28 ± 0.04	18.36 ± 0.03	1.48 ± 0.03	1.32 ± 0.02	5.64 ± 0.06
56.	*Ziziphus xylopyrus* (Kernel)	17.50 ± 0.09	350.10 ± 0.12	241.10 ± 0.13	340.00 ± 0.13	136.00 ± 0.13	44.00 ± 0.08	19.00 ± 0.14	1.50 ± 0.01	11.60 ± 0.12
57.	*Xylia xylocarpa* (Kernel)	33.06 ± 0.11	1836.48 ± 2.16	184.16 ± 0.16	224.30 ± 0.21	115.16 ± 0.10	31.04 ± 0.06	3.20 ± 0.01	0.68 ± 0.01	2.21 ± 0.03

[a] All the values are means of triplicate determinations expressed on dry weight basis.

± Denotes standard error.

TABLE 5.17 Total Soluble Carbohydrates, Starch, Tribal Free Amino Acids, and Vitamins (Niacin and Ascorbic Acid) Content of Edible Piths and Apical Meristem[a]

Sl. No.	Botanical Name Parts Used	Total soluble carbohydrates g $100g^{-1}$	Starch g $100g^{-1}$	Free amino Acid g $100g^{-1}$	Niacin mg $100g^{-1}$	Ascorbic acid mg 100 g^{-1}
1.	*Arenga wightii* (Pith)	12.47 ± 0.18	21.82 ± 0.20	0.05 ± 0.01	21.64 ± 0.06	42.72 ± 0.44
2.	*Caryota urens* (Pith)	0.65 ± 0.43	73.27 ± 0.67	0.03 ± 0.02	70.77 ± 0.05	29.00 ± 0.19
3.	*Phoenix pusilla* (Apical meristem)	3.39 ± 0.19	13.16 ± 0.18	6.32 ± 0.04	32.36 ± 0.02	11.37 ± 0.22

[a] all the values are means of triplicate determinations expressed on dry weight basis
± denotes standard error.

TABLE 5.18 Total Soluble Carbohydrates, Starch, Total Free Amino Acids, and Vitamins (Niacin and Ascorbic acid) Content of Edible Tubers, Rhizomes, Corms and Root Types

Sl. No.	Name of the Plant	Total Soluble Carbohydrates g $100g^{-1}$	Starch g $100g^{-1}$	Total Free amino acid g $100g^{-1}$	Niacin mg $100g^{-1}$	Ascorbic acid mg $100g^{-1}$
1.	*Abelmoschus moschatus* (Root)	15.84 ± 0.08	19.21 ± 0.30	0.09 ± 0.01	58.24 ± 0.28	55.43 ± 0.60
2.	*Amorphophallus paeoniifolius* var. *Campanulatus*	1.28 ± 0.02	51.26 ± 0.14	6.57 ± 0.13	38.06 ± 0.26	32.06 ± 0.21
3.	*Amorphophallus sylvaticus* (Corm)	7.27 ± 0.15	25.45 ± 0.26	0.65 ± 0.06	68.70 ± 0.41	50.64 ± 0.19
4.	*Aponogeton natans* (Tubers)	3.26 ± 0.06	21.16 ± 0.31	1.56 ± 0.08	22.06 ± 0.24	18.26 ± 0.17
5.	*Alocasia macrorhiza* (Tubers)	3.12 ± 0.11	56.25 ± 0.34	1.10 ± 0.12	28.30 ± 0.26	18.26 ± 0.14
6.	*Argyreia pilosa* (Root)	22.20 ± 0.57	52.28 ± 0.37	0.40 ± 0.02	53.12 ± 0.08	68.08 ± 0.11
7.	*Asparagus racemosus* (Tubers)	41.05 ± 0.50	25.35 ± 0.25	0.44 ± 0.06	170.66 ± 0.32	45.79 ± 0.15
8.	*Borassus flabellifer* (Root)	2.26 ± 0.06	33.21 ± 0.36	2.22 ± 0.07	26.16 ± 0.24	24.08 ± 0.13
9.	*Canna indica* (Rhizome)	1.34 ± 0.07	36.00 ± 0.14	9.09 ± 0.11	24.06 ± 0.16	18.21 ± 0.08
10.	*Cissus vitiginea* (Tubers)	0.94 ± 0.02	27.94 ± 0.03	0.17 ± 0.01	15.29 ± 0.12	22.95 ± 0.11
11.	*Colocasia esculenta* (Corm)	11.57 ± 0.28	42.91 ± 0.30	0.12 ± 0.03	88.42 ± 0.17	163.40 ± 0.24
12.	Costus speciosus	1.86 ± 0.06	18.23 ± 0.21	1.02 ± 0.03	26.72 ± 0.18	18.23 ± 0.11
13.	*Curculigo orchioides* (Tubers)	2.55 ± 0.30	36.47 ± 0.78	0.13 ± 0.06	123.85 ± 0.10	14.43 ± 0.13
14.	*Curcuma neilgherrensis.* (Rhizome)	1.76 ± 0.15	38.25 ± 0.45	0.22 ± 0.02	94.23 ± 0.12	86.67 ± 0.34
15.	*Cycas circinalis*. (Tubers)	4.11 ± 0.03	7.94 ± 0.03	0.19 ± 0.01	20.00 ± 0.10	18.36 ± 0.12
16.	*Cyphostemma setosum* (Tubers)	5.58 ± 0.04	7.64 ± 0.06	0.18 ± 0.01	18.82 ± 0.07	20.33 ± 0.12
17.	*Decalepis hamiltonii* (Tubers)	2.96 ± 0.03	12.35 ± 0.03	0.48 ± 0.01	11.76 ± 0.12	30.16 ± 0.10
18.	*Dioscorea alata* (Tubers)	1.29 ± 0.11	64.29 ± 2.23	1.19 ± 0.10	18.06 ± 0.21	15.26 ± 0.16
19.	*Dioscorea bulbifera* var. *vera* (Tubers)	3.46 ± 0.23	18.10 ± 0.17	0.24 ± 0.03	23.69 ± 0.35	106.52 ± 0.11

TABLE 5.18 *(Continued)*

Sl. No.	Name of the Plant	Total Soluble Carbohydrates g $100g^{-1}$	Starch g $100g^{-1}$	Total Free amino acid g $100g^{-1}$	Niacin mg $100g^{-1}$	Ascorbic acid mg $100g^{-1}$
20.	*Dioscorea esculenta* (Tubers)	1.12 ± 0.04	60.60 ± 0.48	5.03 ± 0.12	21.16 ± 0.18	27.64 ± 0.26
21.	*Dioscorea hispida* (Tubers)	2.64 ± 0.03	52.41 ± 0.76	0.86 ± 0.09	24.32 ± 0.14	26.15 ± 0.13
22.	*Dioscorea oppositifolia* var. *dukhumensis* (Tubers)	3.55 ± 0.40	48.13 ± 0.13	0.23 ± 0.03	17.64 ± 0.21	104.79 ± 0.31
23.	*Dioscorea oppositifolia* var. *oppositifolia* (Tubers)	9.73 ± 0.29	40.37 ± 0.46	0.28 ± 0.05	64.55 ± 0.12	80.57 ± 0.12
24.	*Dioscorea pentaphylla* var. *pentaphylla* (Tubers)	5.90 ± 0.11	42.58 ± 0.31	0.13 ± 0.01	53.51 ± 0.27	91.65 ± 0.38
25.	*Dioscorea tomentosa* (Tubers)	3.59 ± 0.30	49.86 ± 0.76	0.25 ± 0.04	88.36 ± 0.12	55.68 ± 0.44
26.	*Dioscorea spicata* (Tubers)	4.06 ± 0.16	52.36 ± 0.33	1.24 ± 0.06	52.06 ± 0.21	33.08 ± 0.17
27.	*Dioscorea wallichii* (Tubers)	1.22 ± 0.09	56.25 ± 0.14	3.57 ± 0.11	66.26 ± 0.18	38.20 ± 0.32
28.	*Dolichos trilobus* (Tubers)	8.79 ± 0.14	32.66 ± 0.28	0.44 ± 0.02	109.99 ± 0.27	57.28 ± 0.56
29.	*Hemidesmus indicus* var. *indicus* (Root)	5.29 ± 0.02	4.70 ± 0.01	0.38 ± 0.02	12.94 ± 0.09	10.49 ± 0.10
30	*Hemidesmus indicus* var. *pubescens* (Root)	6.06 ± 0.03	5.58 ± 0.03	0.94 ± 0.03	11.33 ± 0.07	13.26 ± 0.12
31	*Ipomoea staphylina* (Root)	5.95 ± 0.38	14.79 ± 0.66	0.36 ± 0.02	47.72 ± 0.18	75.47 ± 0.15
32.	*Kedrostis foetidissima* (Tubers)	3.41 ± 0.01	29.41 ± 0.10	0.43 ± 0.08	18.82 ± 0.06	7.87 ± 0.02
33.	*Maerua oblongifolia* (Tubers)	5.88 ± 0.04	4.70 ± 0.03	0.21 ± 0.03	14.12 ± 0.08	21.64 ± 0.06
34.	*Manihot esculenta* (Tubers)	0.44 ± 0.10	64.29 ± 1.38	1.24 ± 0.08	16.08 ± 0.04	13.02 ± 0.03
35	*Maranta arundinacea* (Rhizome)	1.24 ± 0.03	62.31 ± 0.36	3.27 ± 0.21	4.33 ± 0.03	7.37 ± 0.06
36.	*Momordica dioica* (Tubers)	5.88 ± 0.04	22.05 ± 0.06	0.27 ± 0.05	17.65 ± 0.04	3.56 ± 0.02

TABLE 5.18 *(Continued)*

Sl. No.	Name of the Plant	Total Soluble Carbohydrates g $100g^{-1}$	Starch g $100g^{-1}$	Total Free amino acid g $100g^{-1}$	Niacin mg $100g^{-1}$	Ascorbic acid mg $100g^{-1}$
37.	*Nephrolepis auriculata* (Tubers)	1.84 ± 0.06	10.66 ± 0.08	0.33 ± 0.02	12.36 ± 0.06	8.15 ± 0.12
38.	*Nymphaea pubescens* (Tubers)	1.26 ± 0.02	9.74 ± 0.07	0.66 ± 0.03	10.01 ± 0.21	21.06 ± 0.26
39.	*Nymphaea rubra* (Tubers)	2.26 ± 0.03	21.32 ± 0.13	0.48 ± 0.02	6.87 ± 0.12	18.32 ± 0.13
40.	*Parthenocissus neilgherriensis* (Tubers)	3.52 ± 0.13	27.94 ± 0.05	0.18 ± 0.01	18.82 ± 0.02	17.05 ± 0.03
41.	*Sarcostemma acidum* (Tubers)	2.06 ± 0.09	29.36 ± 0.11	0.99 ± 0.03	14.16 ± 0.03	16.21 ± 0.05
42.	*Sterculia urens* (Root)	4.65 ± 0.30	17.88 ± 0.18	0.30 ± 0.04	35.31 ± 0.05	38.83 ± 0.13
43.	*Xanthosoma sagittifolium* (Corm)	2.14 ± 0.23	20.93 ± 1.21	4.76 ± 0.26	22.06 ± 0.08	17.14 ± 0.06
44.	*Xanthosoma violaceum* (Corm)	1.21 ± 0.12	59.34 ± 1.06	3.45 ± 0.03	19.36 ± 0.06	20.18 ± 0.08

[a]All the values are means of triplicate determinations expressed on dry weight basis

[±]Denotes standard error.

TABLE 5.19 Total Soluble Carbohydrates, Starch, Total Free Amino Acid, Vitamins (Niacin and Ascorbic Acid) Content of Greens [a]

Sl. No.	Name of the Plant	Total Soluble Carbohydrates g $100g^{-1}$	Starch g $100g^{-1}$	Free Amino acid g $100g^{-1}$	Niacin mg $100g^{-1}$	Ascorbic acid mg $100g^{-1}$
1.	*Acacia caesia*	4.0 ± 4 ± 0.11	1.93 ± 0.03	1.29 ± 0.07	15.29 ± 0.41	11.80 ± 0.33
2.	*Acacia grahamii*	3.86 ± 0.08	1.24 ± 0.01	1.11 ± 0.04	13.21 ± 0.23	10.30 ± 0.12
3.	*Achyranthes aspera*	2.24 ± 0.04	2.26 ± 0.03	0.86 ± 0.02	12.08 ± 0.11	11.24 ± 0.08
4.	*Achyranthes bidentata*	3.06 ± 0.09	4.98 ± 0.04	0.45 ± 0.01	14.11 ± 0.21	8.66 ± 0.21
5.	*Allmania nodiflora* var. *angustifolia*	3.21 ± 0.05	2.56 ± 0.02	0.94 ± 0.03	13.06 ± 0.14	13.33 ± 0.11
6.	*Allmania nodiflora* var. *procumbens*	2.86 ± 0.04	1.56 ± 0.01	1.02 ± 0.06	15.20 ± 0.33	12.06 ± 0.06
7.	*Aloe vera*	2.32 ± 0.08	3.38 ± 0.04	1.21 ± 0.08	17.11 ± 0.24	9.32 ± 0.05
8.	*Alternanthera bettzickiana*	3.33 ± 0.05	1.66 ± 0.05	0.76 ± 0.05	14.20 ± 0.16	17.06 ± 0.03
9.	*Alternanthera sessilis*	3.52 ± 0.06	1.84 ± 0.06	0.78 ± 0.04	28.32 ± 0.13	15.08 ± 0.08
10.	*Amaranthus roxburghianus*	1.28 ± 0.04	3.48 ± 0.11	0.88 ± 0.05	16.01 ± 0.14	16.24 ± 0.11
11.	*Amaranthus spinosus*	3.31 ± 0.06	4.04 ± 0.04	1.13 ± 0.06	17.05 ± 0.21	9.18 ± 0.07
12.	*Amaranthus tricolor*	2.26 ± 0.03	3.26 ± 0.05	0.94 ± 0.04	18.26 ± 0.09	11.20 ± 0.12
13.	*Amaranthus viridis*	3.01 ± 0.04	3.84 ± 0.04	0.76 ± 0.03	21.11 ± 0.04	9.24 ± 0.06
14.	*Amorphophallus sylvaticus*	3.21 ± 0.05	2.94 ± 0.11	1.11 ± 0.08	20.08 ± 0.06	14.20 ± 0.04
15.	*Asystasia gangetica*	2.33 ± 0.07	4.69 ± 0.15	0.67 ± 0.03	14.12 ± 0.19	13.90 ± 0.14
16.	*Basella alba*	2.94 ± 0.07	3.34 ± 0.04	0.75 ± 0.01	15.88 ± 0.13	16.26 ± 0.07
17.	*Begonia malabarica*	1.64 ± 0.03	2.88 ± 0.05	1.06 ± 0.05	16.28 ± 0.12	15.12 ± 0.03
18.	*Boerhavia diffusa*	1.21 ± 0.02	3.02 ± 0.12	0.99 ± 0.01	13.26 ± 0.04	17.08 ± 0.01
19.	*Boerhavia erecta*	1.08 ± 0.01	2.06 ± 0.04	1.03 ± 0.11	9.14 ± 0.03	19.26 ± 0.1
20.	*Borassus flabellifer*	2.26 ± 0.03	2.14 ± 0.07	0.46 ± 0.03	24.08 ± 0.16	22.64 ± 0.13
21.	*Brassica juncea*	1.35 ± 0.01	2.80 ± 0.13	1.29 ± 0.04	16.46 ± 0.11	18.08 ± 0.09

TABLE 5.19 *(Continued)*

Sl. No.	Name of the Plant	Total Soluble Carbohydrates g $100g^{-1}$	Starch g $100g^{-1}$	Free Amino acid g $100g^{-1}$	Niacin mg $100g^{-1}$	Ascorbic acid mg $100g^{-1}$
22.	*Canthium parvifolium*	1.14 ± 0.02	2.94 ± 0.06	1.14 ± 0.06	18.14 ± 0.10	19.21 ± 0.08
23.	*Cassia obutusifilia*	0.21 ± 0.02	2.36 ± 0.04	0.86 ± 0.03	16.26 ± 0.01	15.38 ± 0.06
24.	*Cassia tora*	1.36 ± 0.04	3.48 ± 0.03	0.54 ± 0.01	18.18 ± 0.08	9.58 ± 0.03
25.	*Capsium annuum*	0.98 ± 0.11	1.78 ± 0.01	0.75 ± 0.02	12.94 ± 0.31	7.34 ± 0.03
26.	*Capsicum frutescens*	1.02 ± 0.01	2.26 ± 0.02	0.33 ± 0.01	24.32 ± 0.04	11.42 ± 0.04
27.	*Cardiospermum canescens*	2.22 ± 0.03	3.26 ± 0.05	0.95 ± 0.02	21.14 ± 0.03	14.38 ± 0.06
28.	*Cardiospermum helicacabum*	1.58 ± 0.05	3.88 ± 0.03	0.94 ± 0.03	22.48 ± 0.05	20.56 ± 0.14
29.	*Cardiospermum microcarpa*	1.66 ± 0.06	3.56 ± 0.02	0.88 ± 0.04	16.38 ± 0.11	18.26 ± 0.09
30	*Celosia argentea*	2.02 ± 0.01	2.88 ± 0.06	0.32 ± 0.01	15.68 ± 0.04	9.14 ± 0.05
31	*Centella asiatica*	1.98 ± 0.03	4.38 ± 0.04	0.56 ± 0.01	18.26 ± 0.12	11.32 ± 0.07
32.	*Cissus quadrangularis*	2.18 ± 0.04	3.98 ± 0.03	1.12 ± 0.04	19.08 ± 0.09	15.68 ± 0.08
33.	*Cissus vitiginea*	1.88 ± 0.02	3.54 ± 0.02	1.02 ± 0.03	21.28 ± 0.06	20.16 ± 0.11
34.	*Cleome gynandra*	2.28 ± 0.05	2.68 ± 0.01	0.56 ± 0.02	24.21 ± 0.08	19.56 ± 0.09
35	*Cleome viscosa*	1.26 ± 0.03	3.33 ± 0.02	0.98 ± 0.02	7.48 ± 0.06	11.34 ± 0.08
36.	*Coccinia grandis*	1.24 ± 0.01	3.24 ± 0.01	1.01 ± 0.03	9.48 ± 0.05	14.18 ± 0.05
37.	*Cocculus hirsutus*	1.36 ± 0.02	2.22 ± 0.03	0.88 ± 0.01	13.08 ± 0.06	13.26 ± 0.06
38.	*Colocasia esculenta*	2.21 ± 0.01	4.43 ± 0.04	1.24 ± 0.02	12.48 ± 0.08	15.06 ± 0.08
39.	*Commelina benghalensis*	5.43 ± 0.02	6.56 ± 0.06	1.14 ± 0.02	11.23 ± 0.11	13.80 ± 0.13
40.	*Commelina ensifolia*	3.33 ± 0.01	3.38 ± 0.05	1.06 ± 0.03	8.24 ± 0.14	10.48 ± 0.06
41.	*Coriandrum sativum*	1.48 ± 0.04	1.98 ± 0.03	1.44 ± 0.02	7.33 ± 0.10	9.24 ± 0.05
42.	*Cycas circinalis*	1.08 ± 0.01	2.24 ± 0.02	1.32 ± 0.04	14.17 ± 0.06	13.46 ± 0.07
43.	*Digera muricata*	2.06 ± 0.03	1.48 ± 0.01	0.92 ± 0.05	15.38 ± 0.05	9.99 ± 0.08

TABLE 5.19 *(Continued)*

Sl. No.	Name of the Plant	Total Soluble Carbohydrates g $100g^{-1}$	Starch g $100g^{-1}$	Free Amino acid g $100g^{-1}$	Niacin mg $100g^{-1}$	Ascorbic acid mg $100g^{-1}$
44.	*Diplocyclos palmatus*	1.63 ± 0.04	2.83 ± 0.03	1.18 ± 0.11	13.64 ± 0.04	11.80 ± 0.07
45.	*Emilia sonchifolia*	1.94 ± 0.05	2.21 ± 0.03	0.48 ± 0.06	16.48 ± 0.08	15.06 ± 0.06
46.	*Euphorbia hirta*	1.44 ± 0.04	1.98 ± 0.05	0.56 ± 0.05	15.14 ± 0.06	19.08 ± 0.09
47.	*Gisekia pharnaceoides*	1.32 ± 0.03	3.48 ± 0.04	0.66 ± 0.04	17.40 ± 0.05	18.14 ± 0.11
48.	*Glinus oppositifolius*	2.12 ± 0.02	2.28 ± 0.03	0.68 ± 0.03	12.56 ± 0.16	10.52 ± 0.08
49.	*Heracleum rigens* var. *rigens*	1.36 ± 0.01	2.32 ± 0.02	0.72 ± 0.02	18.26 ± 0.14	14.33 ± 0.11
50.	*Hybanthus enneaspermus*	2.84 ± 0.06	3.36 ± 0.03	1.11 ± 0.04	21.024 ± 0.08	13.06 ± 0.09
51.	*Ipomoea aquatica*	2.24 ± 0.03	2.92 ± 0.03	0.56 ± 0.02	17.24 ± 0.03	15.24 ± 0.14
52.	*Ipomoea pes-tigridis*	1.94 ± 0.01	2.38 ± 0.02	0.72 ± 0.04	15.18 ± 0.08	16.07 ± 0.08
53.	*Jasminum auriculatum*	2.26 ± 0.02	3.14 ± 0.03	0.88 ± 0.02	14.32 ± 0.11	9.56 ± 0.13
54.	*Jasminum calophyllum*	2.38 ± 0.03	3.06 ± 0.06	0.98 ± 0.01	13.68 ± 0.04	11.38 ± 0.09
55.	*Kalanchoe pinnata*	3.06 ± 0.05	3.24 ± 0.05	1.24 ± 0.02	9.38 ± 0.03	10.38 ± 0.07
56.	*Lathyrus sativus*	2.46 ± 0.02	3.86 ± 0.03	1.18 ± 0.01	10.52 ± 0.09	9.66 ± 0.12
57.	*Leucas montana* var. *wightii*	1.02 ± 0.03	2.94 ± 0.01	0.45 ± 0.03	13.52 ± 0.24	8.90 ± 0.17
58.	*Mollugo pentaphylla*	1.52 ± 0.03	2.32 ± 0.04	0.94 ± 0.01	11.36 ± 0.32	12.36 ± 0.11
59.	*Moringa concanensis*	3.26 ± 0.04	3.58 ± 0.08	1.26 ± 0.03	12.50 ± 0.18	13.14 ± 0.05
60.	*Mukia maderaspatana*	2.20 ± 0.01	3.64 ± 0.07	0.80 ± 0.01	14.11 ± 0.07	7.08 ± 0.11
61.	*Murraya koenigii*	2.84 ± 0.01	3.26 ± 0.04	0.94 ± 0.01	15.26 ± 0.04	27.24 ± 0.08
62.	*Murraya paniculata*	2.26 ± 0.03	3.18 ± 0.03	1.11 ± 0.02	13.14 ± 0.03	19.26 ± 0.04
63.	*Oxalis corniculata*	3.96 ± 0.04	4.21 ± 0.02	1.49 ± 0.07	14.21 ± 0.01	29.37 ± 0.33
64.	*Oxalis latifolia*	3.16 ± 0.05	4.11 ± 0.01	1.26 ± 0.03	21.06 ± 0.02	26.11 ± 0.18

TABLE 5.19 *(Continued)*

Sl. No.	Name of the Plant	Total Soluble Carbohydrates g $100g^{-1}$	Starch g $100g^{-1}$	Free Amino acid g $100g^{-1}$	Niacin mg $100g^{-1}$	Ascorbic acid mg $100g^{-1}$
65.	*Peperomia pellucida*	2.92 ± 0.02	3.56 ± 0.04	1.32 ± 0.02	14.24 ± 0.03	24.06 ± 0.24
66.	*Physalis minima var.indica*	1.86 ± 0.01	3.14 ± 0.05	0.96 ± 0.02	12.06 ± 0.04	15.32 ± 0.11
67.	*Portulaca oleracea* var. *oleracea*	1.54 ± 0.03	2.94 ± 0.03	1.14 ± 0.03	11.24 ± 0.02	16.30 ± 0.08
68.	*Portulaca quadrifida*	2.22 ± 0.06	3.20 ± 0.02	1.02 ± 0.04	18.32 ± 0.01	15.18 ± 0.09
69.	*Premna corymbosa*	1.86 ± 0.06	3.54 ± 0.04	0.98 ± 0.02	14.24 ± 0.03	21.06 ± 0.13
70.	*Psilanthus wightianus*	1.34 ± 0.04	2.26 ± 0.04	1.21 ± 0.03	9.24 ± 0.02	14.22 ± 0.11
71.	*Sarcostemma acidum*	2.22 ± 0.05	5.52 ± 0.03	2.26 ± 0.04	10.36 ± 0.04	15.06 ± 0.08
72.	*Sesbania grandiflora*	2.02 ± 0.01	2.26 ± 0.03	1.11 ± 0.01	13.36 ± 0.06	10.52 ± 0.04
73.	*Solanum anguivi* var. *multiflora*	2.12 ± 0.02	2.86 ± 0.05	1.02 ± 0.01	10.84 ± 0.14	13.36 ± 0.03
74.	*Solanum nigrum*	1.70 ± 0.07	2.98 ± 0.07	1.37 ± 0.02	9.64 ± 0.03	12.58 ± 0.04
75.	*Solanum trilobatum .*	1.96 ± 0.04	2.08 ± 0.02	1.21 ± 0.03	11.24 ± 0.11	14.32 ± 0.05
76.	*Tamarindus indica*	1.54 ± 0.03	2.52 ± 0.03	0.98 ± 0.02	15.36 ± 0.09	15.06 ± 0.14
77.	*Tinospora cordifolia*	2.20 ± 0.02	3.33 ± 0.05	1.04 ± 0.03	20.33 ± 0.17	19.38 ± 0.11
78.	*Trianthema portulacastrum*	2.24 ± 0.05	3.56 ± 0.06	1.96 ± 0.01	24.16 ± 0.21	21.26 ± 0.08
79.	*Vigna radiata* var. *sublobata*	2.26 ± 0.06	2.94 ± 0.04	1.28 ± 0.02	18.36 ± 0.18	17.14 ± 0.06
80.	*Vigna trilobata*	1.98 ± 0.03	2.84 ± 0.02	1.32 ± 0.02	15.18 ± 0.14	15.26 ± 0.05

[a] All the values are means of triplicate determinations expressed on dry weight basis.

± Denotes standard error.

TABLE 5.20 Total Soluble Carbohydrates, Starch, Total Free Aminio Acid, and Vitamins (Niacin and Ascorbic Acid) Content of Seeds and Seed Components[a]

Sl. No.	Name of the plant	Total soluble Carbohydrates g 100 g^{-1}	Starch g 100 g^{-1}	Free amino acid g 100 g^{-1}	Niacin mg 100 g^{-1}	Ascorbic acid mg 100 g^{-1}
1.	*Artocarpus heterophyllus* (Kernel)	4.24 ± 0.26	22.16 ± 0.13	0.84 ± 0.04	18.26 ± 0.08	21.06 ± 0.11
2.	*Atylosia scarabaeoides*	8.34 ± 0.21	28.34 ± 0.12	0.55 ± 0.08	68.12 ± 0.12	12.15 ± 0.12
3.	*Bambusa arundinacea*	3.56 ± 0.23	24.08 ± 0.32	0.98 ± 0.05	11.14 ± 0.03	16.38 ± 0.08
4.	*Borassus flabellifer*	2.86 ± 0.18	33.21 ± 0.36	0.68 ± 0.07	15.25 ± 0.04	28.22 ± 0.14
5.	*Bupleurum wightii*.var. *ramosissimum* (Seed)	1.82 ± 0.14	6.86 ± 0.14	0.48 ± 0.06	14.28 ± 0.06	33.08 ± 0.16
6.	*Canarium strictum* (Kernel)	1.50 ± 0.16	33.04 ± 0.60	0.31 ± 0.01	31.37 ± 0.09	93.02 ± 0.49
7.	*Canavalia gladiata* (Seed)	5.61 ± 0.12	5.01 ± 0.16	2.93 ± 0.05	111.42 ± 0.25	67.09 ± 0.10
8.	*Canavalia virosa* (Seed)	3.62 ± 0.08	3.68 ± 0.05	2.21 ± 0.08	48.21 ± 0.17	54.26 ± 0.18
9.	*Capparis zeylanica* (Seed)	1.64 ± 0.04	8.92 ± 0.11	0.56 ± 0.04	17.20 ± 0.06	14.14 ± 0.11
10.	*Celtis philippensis* var. wightii (Seed)	1.36 ± 0.03	5.34 ± 0.06	0.66 ± 0.05	12.36 ± 0.05	16.28 ± 0.08
11.	*Chamaecrista absus* (Seed)	2.86 ± 0.05	3.18 ± 0.03	1.82 ± 0.03	43.06 ± 0.08	33.08 ± 0.15
12.	*Cycas circinalis* (Kernel)	5.08 ± 0.04	58.68 ± 0.14	0.25 ± 0.06	41.25 ± 0.09	9.45 ± 0.06
13.	*Dolichos lablab* var.*vulgaris* (seed)	4.98 ± 0.11	16.24 ± 0.06	0.44 ± 0.03	34.24 ± 0.12	28.26 ± 0.19
14.	*Dolichos trilobus* (Seed)	5.31 ± 0.13	18.11 ± 0.08	0.54 ± 0.05	48.11 ± 0.13	34.18 ± 0.08
15.	*Drypetes sepiaria* (Seed)	2.33 ± 0.10	9.26 ± 0.06	0.88 ± 0.03	22.01 ± 0.14	32.08 ± 0.14
16.	*Elaeocarpus tectorius* (Kernel)	2.73 ± 0.12	8.23 ± 0.11	0.33 ± 0.08	35.31 ± 0.06	61.92 ± 0.21
17.	*Eleusine coracana* (Seed)	3.26 ± 0.16	14.06 ± 0.05	0.44 ± 0.01	16.23 ± 0.05	44.06 ± 0.18
18.	*Ensete superbum* (Seed)	1.58 ± 0.04	9.28 ± 0.03	0.82 ± 0.03	18.16 ± 0.07	58.08 ± 0.38
19.	*Entada rheedi* (Kernel)	11.78 ± 0.32	9.35 ± 0.04	0.87 ± 0.01	27.32 ± 0.19	208.19 ± 0.19
20.	*Givotia rottleriformis* (Kernel)	1.10 ± 0.02	3.57 ± 0.09	0.24 ± 0.25	35.31 ± 0.01	14.33 ± 0.06

TABLE 5.20 *(Continued)*

Sl. No.	Name of the plant	Total soluble Carbohydrates g 100 g^{-1}	Starch g 100 g^{-1}	Free amino acid g 100 g^{-1}	Niacin mg 100 g^{-1}	Ascorbic acid mg 100 g^{-1}
21.	*Heracleum rigens* var. *rigens* (Seed)	2.30 ± 0.08	5.86 ± 0.10	0.73 ± 0.02	21.26 ± 0.11	15.16 ± 0.03
22.	*Impatiens balsamina* (Seed)	2.16 ± 0.06	9.10 ± 0.07	0.66 ± 0.05	18.24 ± 0.09	19.26 ± 0.08
23.	*Lablab purpureus* var. *lignosus* (Seed)	3.09 ± 0.07	16.90 ± 0.13	0.51 ± 0.03	70.55 ± 0.35	32.78 ± 0.14
24.	*Lablab purpureus* var. *purpureus* (Seed)	4.11 ± 0.05	12.08 ± 0.08	1.01 ± 0.02	54.26 ± 0.22	28.05 ± 0.08
25.	*Macrotyloma uniflorum* (seed)	3.26 ± 0.12	14.12 ± 0.14	0.78 ± 0.04	42.16 ± 0.31	38.24 ± 0.26
26.	*Moringa concanensis* (Kernel)	10.35 ± 0.09	3.57 ± 0.19	0.54 ± 0.05	42.22 ± 0.41	214.33 ± 0.22
27.	*Mucuna atropurpurea* (Kernel)	1.24 ± 0.06	21.88 ± 0.24	1.15 ± 0.07	86.32 ± 0.09	75.39 ± 0.43
28.	*Mucuna pruriens* var. *pruriens* (Seed)	2.88 ± 0.03	21.22 ± 0.08	1.08 ± 0.03	44.33 ± 0.11	56.21 ± 0.14
29.	*Mucuna pruriens* var. utilis (White Coloured Seed Coat) (Seed)	3.12 ± 0.04	22.46 ± 0.12	1.21 ± 0.02	48.16 ± 0.08	62.30 ± 0.08
30.	*Mucuna pruriens* var. utilis (Black Coloured Seed Coat) (Seed)	3.46 ± 0.02	24.06 ± 0.09	1.33 ± 0.04	52.38 ± 0.09	58.34 ± 0.12
31.	*Neonotonia wightii*. var. *coimbatorensis* (Seed)	4.12 ± 0.03	18.11 ± 0.13	0.44 ± 0.13	84.12 ± 0.11	84.34 ± 0.12
32.	*Ocimum gratissimum* (Seed)	1.31 ± 0.05	11.23 ± 0.15	0.40 ± 0.06	47.35 ± 0.12	203.04 ± 0.42
33.	*Oryza meyeriana* var. *granulata* (Grain)	2.38 ± 0.11	32.68 ± 0.24	1.02 ± 0.03	33.21 ± 0.08	22.86 ± 0.14
34.	*Pithecellobium dulce* (Aril)	3.38 ± 0.04	12.33 ± 0.16	0.88 ± 0.02	38.38 ± 0.09	32.16 ± 0.08
35.	*Rhynchosia cana* (Seed)	3.50 ± 0.03	27.10 ± 0.54	0.70 ± 0.03	41.38 ± 0.03	76.94 ± 0.18
36.	*Rhynchosia filipes* (Seed)	13.32 ± 0.16	17.62 ± 0.13	0.25 ± 0.01	30.33 ± 0.14	59.09 ± 0.51
37.	*Rhynchosia rufescens* (Seed)	4.56 ± 0.04	24.14 ± 0.11	0.86 ± 0.11	24.11 ± 0.09	56.24 ± 0.08
38.	*Rhynchosia suaveolens* (Seed)	5.38 ± 0.11	16.28 ± 0.08	0.92 ± 0.06	26.32 ± 0.16	48.32 ± 0.12
39.	*Sapindus emarginatus* (Seed)	3.48 ± 0.04	12.10 ± 0.06	0.82 ± 0.05	23.06 ± 0.07	33.21 ± 0.08

TABLE 5.20 *(Continued)*

Sl. No.	Name of the plant	Total soluble Carbohydrates g 100 g^{-1}	Starch g 100 g^{-1}	Free amino acid g 100 g^{-1}	Niacin mg 100 g^{-1}	Ascorbic acid mg 100 g^{-1}
40.	*Sesamum indicum* (Seed)	1.03 ± 0.17	2.70 ± 0.12	0.15 ± 0.16	26.13 ± 0.11	34.39 ± 0.32
41.	*Shorea roxburghii* (Cotyledon)	2.22 ± 0.06	16.36 ± 0.18	0.74 ± 0.03	6.88 ± 0.09	44.30 ± 0.26
42.	*Sterculia foetida* (Kernel)	6.31 ± 0.03	7.38 ± 0.21	0.30 ± 0.03	5.84 ± 0.13	95.44 ± 0.12
43.	*Sterculia guttata* (Kernel)	5.27 ± 0.01	9.72 ± 0.26	0.51 ± 0.09	58.84 ± 0.16	60.35 ± 0.14
44.	*Sterculia urens* (Kernel)	7.05 ± 0.09	25.68 ± 0.21	0.11 ± 0.15	2.99 ± 0.06	29.36 ± 0.09
45.	*Strychnos nux-vomica* (Kernel)	1.98 ± 0.03	26.85 ± 0.21	0.14 ± 0.07	73.58 ± 0.03	27.37 ± 0.19
46.	*Tamarindus indica* (Kernel)	6.45 ± 0.13	12.66 ± 0.04	0.50 ± 0.08	76.67 ± 0.09	50.42 ± 0.17
47.	*Teramnus labialis* (Seed)	4.60 ± 0.09	9.97 ± 0.23	0.89 ± 0.02	53.00 ± 0.02	21.25 ± 0.17
48.	*Terminalia bellirica* (Kernel)	4.69 ± 0.15	2.36 ± 0.01	0.33 ± 0.01	11.73 ± 0.11	34.53 ± 0.41
49.	*Terminalia chebula* (Kernel)	3.31 ± 0.07	3.86 ± 0.04	0.38 ± 0.03	22.08 ± 0.09	32.14 ± 0.26
50.	*Vigna bourneae* (Seed)	4.11 ± 0.03	25.36 ± 0.08	0.98 ± 0.05	26.18 ± 0.06	28.36 ± 0.14
51.	*Vigna radiata* (Seed)	4.38 ± 0.11	13.14 ± 0.11	1.24 ± 0.01	38.14 ± 0.14	64.38 ± 0.13
52.	*Vigna trilobata* (Seed)	4.88 ± 0.02	15.77 ± 0.27	0.24 ± 0.02	38.26 ± 0.02	68.09 ± 0.52
53.	*Vigna unguiculata subsp. cylindrica* (Seed)	3.07 ± 0.09	22.16 ± 0.12	0.85 ± 0.02	14.62 ± 0.15	127.67 ± 0.14
54.	*Vigna unguiculata* subsp. *unguiculata* (Seed)	4.39 ± 0.05	21.42 ± 0.17	1.44 ± 0.06	3.88 ± 0.17	49.19 ± 0.55
55.	*Ziziphus rugosa* (Kernel)	2.98 ± 0.04	18.38 ± 0.11	0.92 ± 0.04	32.26 ± 0.6	38.24 ± 0.38
56.	*Ziziphus xylopyrus* (Kernel)	3.58 ± 0.07	3.31 ± 0.26	0.16 ± 0.02	94.64 ± 0.21	92.30 ± 0.50
57.	*Xylia xylocarpa* (Kernel)	3.66 ± 0.04	14.11 ± 0.08	0.33 ± 0.05	44.12 ± 0.16	33.68 ± 0.30

[a] all the values are means of triplicate determinations expressed on dry weight basis
± denotes standard error

The total soluble carbohydrates of the investigated samples ranged between 0.65% and 41.05%. Total soluble carbohydrates content of tuber of *Asparagus racemosus,* roots of *Argyreia pilosa* and *Abelmoschus moschatus* contain more than 15%. The examination that tubers of *Asparagus racemosus* contain more than 40% of total soluble carbohydrates is noteworthy.

The starch content of the investigated samples is ranged between 1.48% and 73.27%. The tubers of *Alocasia macrorhiza, Dioscorea alata, D. esculenta, D. hispida, D. spicata, D. wallichi, Manicot esculenta,* corms of *Amorphophallus paeoniifolius* var. *campanulatus, Xanthosoma violeceum,* rhizome of *Maranta arundinacea*, root of *Argyreia pilosa*, pith of *Caryota urens* and seeds of *Cycas circinalis* exhibited more than 50% of starch.

The contents of total free amino acids of the apical meristem of *Phoenix pusilla*, corms of *Amorphophallus paeoniifolius* var. *campanulatus*, the rhizome of *Canna indica* and tuber of *Dioscorea esculenta* exhibit more than 5%.

5.1.3.2 VITAMINS (NIACIN AND ASCORBIC ACID)

The contents of niacin and ascorbic acid are furnished in Tables 5.17–5.20. Relatively high level of niacin is found in the tubers of *Asparagus racemosus*, *Curculigo orchioides*, and *Dolichos trilobus* and in the seed of *Canavalia gladiata*. The high content of ascorbic acid is found in the tubers of *Dioscorea bulbifera* var *vera*, *D. oppositifolia* var *dukhumensis*, corm of *Colocasia esculenta*, Kernel of *Entada rheedi*, *Moringa concanensis*, seeds of *Ocimum gratissimum* and *Vigna unguiculata* subsp.*cylindrica.*

5.2 ANTINUTRITIONAL FACTORS

To continue to exist in adverse growing conditions, plants commonly amalgamate a range of secondary metabolites as a part of their protection against attack by herbivores, insects, and pathogens. If animals or humans consume these plants, these compounds may cause undesirable physiological effects. The terms antinutrient or natural toxicant have been widely utilized to explain plant resistance metabolites in food and nutrition text.

Wild edible plants are rich in numerous nutrients. Normally, rural people consume these plants. However, the main problem related to the nutritional exploitation of these kinds of plants is the presence of antinutritional and toxic principles. The antinutritional factors may be characterized as those materials made in natural feedstuffs by the normal metabolism of the species and also by many mechanisms. In addition to this, these factors take advantage of outcomes contrary to optimum nutrition (Kumar, 1983).

In a developing country like India, traditionally wild edible fruits are the only fruits consumed by rural people. The native fruits collected from wild edible plants play a major role in the food and nutrient security of rural poor in general and tribal mass in particular. As a result, in recent years, a growing interest has emerged to estimate various wild edible fruits for their nutritional value. On the other hand, some wild fruits have nutritional as well as antinutritional properties. These properties adversely affect human health. Keeping this in mind the need for evaluation of antinutritional properties of these wild fruits is necessary so that the knowledge derived can be used to support adequate consumption for the human body.

The antinutritional factors such as phytic acid, tannin, saponin, oxalic acid, badly affect the nutrients needed by the body. Further, this inhibits protein digestion, growth, and iron and zinc absorption. Oxalic acid binds to calcium, and this is found naturally in a variety of fruits, vegetables, nuts, grains, and legumes. It has also been reported that oxalates causes irritation and swelling in the mouth and throat. Phytate is an organic acid as well as a major component of plant storage organs which serves as phosphate source for germination and growth. It decreases calcium bioavailability and forms calcium phytate complexes that inhibit the absorption of Fe, Zn. Tannins have a capacity to precipitate to certain protein, and it forms a complex that inhibits the digestibility and palatability. Saponin is a natural glycoside that generally believed to be non-poisonous to warm-blooded animals, but they are dangerous when injected into the bloodstream and quickly hemolyzed red blood cells.

Wild edible plants are the major source of minerals, vitamins, and fiber, and also provide essential nutrients for human health. The nutritional value of many wild edible plants found to be comparable with cultivated commercial fruits. On the other hand, many wild edible plants are reported to have antinutritional properties like oxalate, phytate, tannin, and saponin. Though sporadic reports on levels of antinutritional factors of some underutilized and unexplored wild edible plants are available, yet, scientific information of these wild edible plants regarding their antinutritional properties is lacking especially in case of the wild edible plants.

The data on antinutritional factors, total free phenolics, tannins, L-Dopa and hydrogen cyanide are presented in Tables 5.21–5.25.

5.2.1 TOTAL FREE PHENOLICS

In the investigated samples, the content of total free phenolics ranged from 0.03% to 4.63%. Rhizome of *Curcuma neilgherrensis*, tubers of *Dioscorea bulbifera* var. *vera, D. hispida* corm of *Amorphophallus paeoniifolius* var. *campanulatus*, kernel of *Entada rheedi, Moringa concanensis, Mucuna atropurpurea, Tamarindus indica, Xylia xylocarpa* seeds of *Canavalia gladiata, C. virosa, Dolichos trilobus, Mucuna pruriens* var. *pruriens*, *Mucuna pruriens* var. *utilis* (black colored seed coat), *Mucuna pruriens* var. *utilis*, (white colored seed coat) *Neonotonia wightii* var. *coimbatorensis, Rhynchosia cana, R.rufescens, R. suaveolens, Teramnus labialis, Vigna bourneae, V. radiata* var. *sublobata, V. trilobata, V. ungiculata* subsp. *cylindrica* and *V. unguiculata* subsp. *ungiculata,* the total free phenolics content is found to be more than 1%.

5.2.2 TANNINS

The tannin content of all the investigated samples ranged from 0.02% to 1.55%. In the tubers of *Dioscorea bulbifera* var. *vera* and corm of *Amorphophallus paeonifolius* var. *campanulatus*, the tannin content is found to be above 1%. In all the other investigated samples tannin content is found to be less than 1%.

5.2.3 L-DOPA

Relatively high levels of L-Dopa were recorded in the seeds of *Atylosia scarabaeoides, Canavalia gladiata, Mucuna pruriens* var. *pruriens*, *M. pruriens* var. *utilis* (black and white colored seed coat), kernels of *Entada rheedi,* and *Tamarindus indica.*

In the investigated plant samples hydrogen cyanide content is found to be negligible (below 0.71 mg 100 g^{-1}).

TABLE 5.21 Antinutritional Factors of Edible Pith and Apical Meristem [a]

Sl. No.	Botanical Name	Total Free Phenolic g 100 g^{-1}	Tannim g 100 g^{-1}	Hydrogen Cyanide mg 100 g^{-1}
1.	*Arenga wightii* (Pith)	0.05 ± 0.01	Nd	0.46 ± 0.06
2.	*Caryota urens* (Pith)	0.03 ± 0.001	0.04 ± 0.01	0.53 ± 0.03
3.	*Phoenix pusilla* (Apical meristem)	0.16 ± 0.02	0.12 ± 0.03	0.14 ± 0.02

[a] All the values are means of triplicate determinations expressed on dry weight basis.

± Denotes standard error.

ND – not detected.

TABLE 5.22 Antirutifioral Factors of Edible Tubers, Rhizomes, Corms and Root-Types [a]

Sl. No.	Botanical Name	Total Free Phenolic g $100g^{-1}$	Tannin g $100g^{-1}$	Hydrogen Cyanide mg $100g^{-1}$
	Abelmoschus moschatus (Root)	0.24 ± 0.03	0.22 ± 0.02	0.26 ± 0.01
	Amorphophallus paeoniifolius var. campanulatus (Corm)	1.19 ± 0.01	1.34 ± 0.04	0.12 ± 0.02
	Amorphophallus sylvaticus (Corm)	0.69 ± 0.04	0.24 ± 0.01	0.26 ± 0.01
	Aponogeton natans (Tubers)	0.24 ± 0.03	0.11 ± 0.01	0.14 ± 0.01
	Alocasia macrorhiza (Tubers)	0.83 ± 0.03	0.62 ± 0.01	0.13 ± 0.02
	Argyreia pilosa (Root)	0.11 ± 0.02	0.13 ± 0.01	0.48 ± 0.02
	Asparagus racemosus (Tubers)	0.28 ± 0.03	0.31 ± 0.04	0.13 ± 0.01
	Borassus flabellifer (Root)	0.24 ± 0.02	0.18 ± 0.02	0.16 ± 0.01
	Canna indica (Rhizome)	0.50 ± 0.04	0.13 ± 0.01	0.11 ± 0.01
	Cissus vitiginea (Tubers)	0.12 ± 0.02	0.14 ± 0.01	0.15 ± 0.01
	Colocasia esculenta (Corm)	0.19 ± 0.03	0.18 ± 0.01	0.20 ± 0.03
	Costus speciosus (Rhizome)	0.29 ± 0.02	0.18 ± 0.02	0.11 ± 0.01
	Curculigo orchioides (Tubers)	0.39 ± 0.03	0.08 ± 0.01	0.33 ± 0.04
	Curcuma neilgherrensis. (Rhizome)	1.73 ± 0.11	0.19 ± 0.02	0.31 ± 0.03
	Cycas circinalis (Tubers)	0.20 ± 0.05	0.12 ± 0.01	0.11 ± 0.01
	Cyphostemma setosum (Tubers)	0.41 ± 0.06	0.34 ± 0.05	0.13 ± 0.01
	Decalepis hamiltonii (Tubers)	0.27 ± 0.05	0.12 ± 0.01	0.14 ± 0.01
	Dioscorea alata (Tubers)	0.49 ± 0.03	0.51 ± 0.02	0.15 ± 0.02
	Dioscorea bulbifera var. *vera* (Tubers)	3.37 ± 0.15	1.55 ± 0.06	0.17 ± 0.02
	Dioscorea esculenta (Tubers)	0.79 ± 0.07	0.20 ± 0.01	0.21 ± 0.03
	Dioscorea hispida (Tubers)	1.21 ± 0.04	0.28 ± 0.06	0.15 ± 0.01
	Dioscorea oppositifolia var. *dukhumensis* (Tubers)	0.24 ± 0.01	0.09 ± 0.01	0.09 ± 0.01
	Dioscorea oppositifolia var. *oppositifolia* (Tubers)	0.34 ± 0.02	0.11 ± 0.01	0.30 ± 0.02

TABLE 5.22 *(Continued)*

Sl. No.	Botanical Name	Total Free Phenolic g $100g^{-1}$	Tannin g $100g^{-1}$	Hydrogen Cyanide mg $100g^{-1}$
1.	*Dioscorea pentaphylla* var. *pentaphylla* (Tubers)	0.31 ± 0.03	0.14 ± 0.02	0.17 ± 0.01
2.	*Dioscorea tomentosa* (Tubers)	0.44 ± 0.02	0.20 ± 0.03	0.10 ± 0.01
3.	*Dioscorea spicata* (Tubers)	0.26 ± 0.01	0.10 ± 0.01	0.18 ± 0.02
4.	*Dioscorea wallichii* (Tubers)	0.33 ± 0.02	0.14 ± 0.01	0.16 ± 0.03
5.	*Dolichos trilobus* (Tubers)	0.18 ± 0.02	0.12 ± 0.02	0.32 ± 0.07
6.	*Hemidesmus indicus* var. *indicus* (Root)	0.24 ± 0.03	0.08 ± 0.01	0.15 ± 0.01
7.	*Hemidesmus indicus* var. *pubescens* (Root)	0.21 ± 0.02	0.11 ± 0.02	0.08 ± 0.01
8.	*Ipomoea staphylina* (Root)	0.35 ± 0.03	0.12 ± 0.01	0.53 ± 0.06
9.	*Kedrostis foetidissima* (Tubers)	0.12 ± 0.01	0.08 ± 0.01	0.08 ± 0.01
10.	*Maerua oblongifolia* (Tubers)	0.16 ± 0.02	0.08 ± 0.01	0.09 ± 0.02
11.	*Manihot esculenta* (Tubers)	0.46 ± 0.02	0.14 ± 0.02	0.29 ± 0.01
12.	*Maranta arundinacea* (Rhizome)	0.09 ± 0.01	0.26 ± 0.01	0.14 ± 0.02
13.	*Momordica dioica* (Tubers)	0.38 ± 0.02	0.32 ± 0.04	0.11 ± 0.01
14.	*Nephrolepis auriculata* (Tubers)	0.28 ± 0.02	0.18 ± 0.01	0.19 ± 0.02
15.	*Nymphaea pubescens* (Tubers)	0.21 ± 0.05	0.18 ± 0.02	0.12 ± 0.01
16.	*Nymphaea rubra* (Tubers)	0.20 ± 0.01	0.11 ± 0.03	0.14 ± 0.01
17.	*Parthenocissus neilgherriensis* (Tubers)	0.48 ± 0.02	0.13 ± 0.02	0.08 ± 0.01
18.	*Sarcostemma acidum* (Tubers)	0.38 ± 0.06	0.26 ± 0.02	0.14 ± 0.02
19.	*Sterculia urens* (Root)	0.80 ± 0.03	1.01 ± 0.04	0.62 ± 0.04
20.	*Xanthosoma sagittifolium* (Corm)	0.16 ± 0.002	0.13 ± 0.01	0.11 ± 0.01
21.	*Xanthosoma violaceum* (Corm)	0.13 ± 0.001	0.31 ± 0.04	0.09 ± 0.01

[a] All the values are means of triplicate determinations expressed on dry weight basis.

± Denotes standard error.

TABLE 5.23 Antinutritional Factors of Greens[a]

Sl. No	Name of the Plant	Total Free Phenolic g 100 g^{-1}	Tannin g 100 g^{-1}	Hydrogen Cyanide g 100 g^{-1}
1.	*Acacia caesia*	0.96 ± 0.13	0.02 ± 0.001	0.07 ± 0.03
2.	*Acacia grahamii*	0.34 ± 0.03	0.14 ± 0.01	0.06 ± 0.01
3.	*Achyranthes aspera*	0.28 ± 0.01	0.08 ± 0.01	0.09 ± 0.02
4.	*Achyranthes bidentata*	0.42 ± 0.03	0.04 ± 0.001	0.04 ± 0.01
5.	*Allmania nodiflora* var. *angustifolia*	0.68 ± 0.06	0.12 ± 0.01	0.08 ± 0.01
6.	*Allmania nodiflora* var. *procumbens.*	0.72 ± 0.08	0.08 ± 0.01	0.04 ± 0.01
7.	*Aloe vera*	0.58 ± 0.06	0.11 ± 0.02	0.06 ± 0.01
8.	*Alternanthera bettzickiana*	0.38 ± 0.04	0.08 ± 0.01	0.03 ± 0.001
9.	*Alternanthera sessilis*	0.46 ± 0.03	0.14 ± 0.02	0.06 ± 0.01
10.	*Amaranthus roxburghianus*	0.38 ± 0.01	0.13 ± 0.03	0.03 ± 0.001
11.	*Amaranthus spinosus*	0.52 ± 0.01	0.03 ± 0.001	0.04 ± 0.01
12.	*Amaranthus tricolor*	0.46 ± 0.02	0.08 ± 0.01	0.06 ± 0.01
13.	*Amaranthus viridis.*	0.56 ± 0.04	0.09 ± 0.01	0.04 ± 0.01
14.	*Amorphophallus sylvaticus*	0.52 ± 0.03	0.06 ± 0.01	0.06 ± 0.02
15.	*Asystasia gangetica*	0.53 ± 0.02	0.21 ± 0.03	0.07 ± 0.02
16.	*Basella alba*	0.44 ± 0.07	0.52 ± 0.04	0.06 ± 0.001
17.	*Begonia malabarica*	0.76 ± 0.09	0.31 ± 0.06	0.12 ± 0.01
18.	*Borahavia diffusa*	0.52 ± 0.06	0.12 ± 0.02	0.08 ± 0.01
19.	*Boerhavia erecta*	0.49 ± 0.08	0.16 ± 0.01	0.06 ± 0.001
20.	*Borassus flabellifer*	0.32 ± 0.04	0.11 ± 0.02	0.04 ± 0.001
21.	*Brassica juncea*	0.66 ± 0.08	0.14 ± 0.001	0.06 ± 0.01
22.	*Canthium parvifolium*	0.51 ± 0.06	0.08 ± 0.01	0.04 ± 0.01
23.	*Cassia obtusifolia*	0.50 ± 0.06	0.14 ± 0.02	0.06 ± 0.01
24.	*Cassia tora*	0.66 ± 0.03	0.17 ± 0.03	0.09 ± 0.001
25.	*Capsium annuum*	0.60 ± 0.08	0.10 ± 0.01	0.03 ± 0.001
26.	*Capsicum frutescens*	0.52 ± 0.04	0.12 ± 0.02	0.10 ± 0.01
27.	*Cardiospermum canescens*	0.58 ± 0.03	0.11 ± 0.01	0.08 ± 0.044
28.	*Cardiospermum helicacabum*	0.62 ± 0.02	0.18 ± 0.02	0.06 ± 0.001
29.	*Cardiospermum microcarpa*	0.46 ± 0.03	0.08 ± 0.01	0.11 ± 0.01
30	*Celosia argentea*	0.44 ± 0.02	0.12 ± 0.04	0.09 ± 0.01
31	*Centella asiatica*	0.64 ± 0.03	0.11 ± 0.01	0.14 ± 0.02
32.	*Cissus quadrangularis*	0.36 ± 0.02	0.14 ± 0.03	0.12 ± 0.01
33.	*Cissus vitiginea*	0.41 ± 0.06	0.18 ± 0.02	0.13 ± 0.02

TABLE 5.23 *(Continued)*

Sl. No	Name of the Plant	Total Free Phenolic g 100 g^{-1}	Tannin g 100 g^{-1}	Hydrogen Cyanide g 100 g^{-1}
34.	*Cleome gynandra*	0.33 ± 0.04	0.06 ± 0.001	0.11 ± 0.01
35	*Cleome viscosa*	0.48 ± 0.09	0.11 ± 0.01	0.08 ± 0.001
36.	*Coccinia grandis*	0.24 ± 0.04	0.08 ± 0.02	0.06 ± 0.001
37.	*Cocculus hirsutus*	0.36 ± 0.03	0.14 ± 0.01	0.09 ± 0.01
38.	*Colocasia esculenta*	0.38 ± 0.02	0.16 ± 0.02	0.12 ± 0.02
39.	*Commelina benghalensis*	0.98 ± 0.10	0.16 ± 0.01	0.06 ± 0.01
40.	*Commelina ensifolia*	0.56 ± 0.08	0.11 ± 0.02	0.08 ± 0.001
41.	*Coriandrum sativum*	0.33 ± 0.04	0.08 ± 0.01	0.09 ± 0.01
42.	*Cycas circinalis*	0.46 ± 0.02	0.16 ± 0.02	0.06 ± 0.01
43.	*Digera muricata*	0.76 ± 0.08	0.18 ± 0.03	0.05 ± 0.01
44.	*Diplocyclos palmatus*	0.82 ± 0.04	0.11 ± 0.01	0.03 ± 0.001
45.	*Emilia sonchifolia*	0.52 ± 0.03	0.09 ± 0.01	0.09 ± 0.01
46.	*Euphorbia hirta*	0.63 ± 0.02	0.10 ± 0.01	0.10 ± 0.02
47.	*Gisekia pharnaceoides*	0.81 ± 0.09	0.14 ± 0.01	0.09 ± 0.01
48.	*Glinus oppositifolius*	0.78 ± 0.06	0.11 ± 0.01	0.12 ± 0.02
49.	*Heracleum rigens* var. *regens*	0.62 ± 0.07	0.09 ± 0.04	0.08 ± 0.01
50.	*Hybanthus enneaspermus*	0.56 ± 0.06	0.16 ± 0.03	0.06 ± 0.001
51.	*Ipomoea aquatic*	0.38 ± 0.02	0.09 ± 0.02	0.05 ± 0.001
52.	*Ipomoea pestigridis*	0.42 ± 0.01	0.07 ± 0.01	0.14 ± 0.01
53.	*Jasminum auriculatum*	0.36 ± 0.03	0.19 ± 0.03	0.04 ± 0.002
54.	*Jasminum calophyllum*	0.41 ± 0.02	0.10 ± 0.01	0.06 ± 0.001
55.	*Kalanchoe pinnata*	0.56 ± 0.01	0.08 ± 0.02	0.14 ± 0.02
56.	*Lathyrus sativus*	0.62 ± 0.03	0.21 ± 0.04	0.04 ± 0.001
57.	*Leucas montana* var. *wightii.*	0.66 ± 0.14	0.08 ± 0.01	0.10 ± 0.01
58.	*Mollugo pentaphylla*	0.58 ± 0.03	0.19 ± 0.05	0.08 ± 0.002
59.	*Moringa concanensis*	0.64 ± 0.03	0.21 ± 0.02	0.06 ± 0.01
60.	*Mukia maderaspatana*	0.42 ± 0.06	0.14 ± 0.01	0.08 ± 0.01
61.	*Murraya koenigii*	0.38 ± 0.04	0.16 ± 0.01	0.09 ± 0.02
62.	*Murraya paniculata*	0.32 ± 0.03	0.12 ± 0.02	0.04 ± 0.01
63.	*Oxalis corniculata*	0.47 ± 0.03	0.12 ± 0.02	0.04 ± 0.001
64.	*Oxalis latifolia*	0.55 ± 0.03	0.08 ± 0.01	0.03 ± 0.001
65.	*Peperomia pellucida*	0.62 ± 0.01	0.21 ± 0.02	0.04 ± 0.001
66.	*Physalis minima* var. *indica*	0.66 ± 0.02	0.18 ± 0.01	0.06 ± 0.001
67.	*Portulaca oleracea* var. *oleracea*	0.63 ± 0.03	0.15 ± 0.01	0.11 ± 0.01

TABLE 5.23 *(Continued)*

Sl. No	Name of the Plant	Total Free Phenolic g 100 g^{-1}	Tannin g 100 g^{-1}	Hydrogen Cyanide g 100 g^{-1}
68.	*Portulaca quadrifida*	0.62 ± 0.07	0.03 ± 0.001	0.14 ± 0.01
69.	*Premna corymbosa*	0.58 ± 0.05	0.12 ± 0.01	0.06 ± 0.002
70.	*Psilanthus wightianus*	0.60 ± 0.04	0.08 ± 0.002	0.05 ± 0.001
71.	*Sarcostemma acidum*	1.21 ± 0.06	0.19 ± 0.02	0.08 ± 0.01
72.	*Sesbania grandiflora*	0.76 ± 0.04	0.18 ± 0.01	0.10 ± 0.02
73.	*Solanum anguivi* var. *multiflora*	0.62 ± 0.05	0.16 ± 0.01	0.09 ± 0.01
74.	*Solanum nigrum*	0.98 ± 0.04	0.18 ± 0.02	0.06 ± 0.01
75.	*Solanum trilobatum*	0.58 ± 0.03	0.20 ± 0.01	0.07 ± 0.01
76.	*Tamarindus indica*	1.02 ± 0.05	0.38 ± 0.03	0.06 ± 0.002
77.	*Tinospora cordifolia*	0.56 ± 0.03	0.26 ± 0.02	0.05 ± 0.001
78.	*Trianthema portulacastrum*	0.99 ± 0..02	0.18 ± 0.01	0.03 ± 0.001
79.	*Vigna radiata*	0.40 ± 0.03	0.14 ± 0.02	0.09 ± 0.02
80.	*Vigna trilobata*	0.36 ± 0.02	0.11 ± 0.01	0.08 ± 0.01

[a] All the values are means of triplicate determinations expressed on dry weight basis.

± Denotes standard error.

TABLE 5.24 Antinutritional Factors of Edible Seeds[a]

Sl. No.	Name of the Plant	Components		
		Total Free Phenolic g 100g^{-1}	Tannin g 100g^{-1}	Hydrogen Cyanide mg 100g^{-1}
1.	*Artocarpus heterophyllus* (Kernel)	0.48 ± 0.06	0.14 ± 0.03	0.10 ± 0.01
2.	*Bambusa arundinacea* (Seed)	0.22 ± 0.01	0.10 ± 0.02	0.15 ± 0.04
3.	*Borassus flabellifer* (Endosperm)	0.28 ± 0.03	0.16 ± 0.01	0.14 ± 0.01
4.	*Bupleurum wightii*. var. *ramosissimum* (Seed)	0.52 ± 0.08	0.26 ± 0.04	0.16 ± 0.02
5.	*Canarium strictum* (Kernel)	0.16 ± 0.02	0.21 ± 0.02	0.33 ± 0.07
6.	*Capparis zeylanica* (Seed)	0.18 ± 0.02	0.11 ± 0.03	0.10 ± 0.03
7.	*Celtis philippensis* var. wightii (Seed)	0.21 ± 0.03	0.08 ± 0.01	0.10 ± 0.02
8.	*Cycas circinalis* (Kernel)	0.11 ± 0.01	0.18 ± 0.02	0.20 ± 0.03
9.	*Drypetes sepiaria* (Seed)	0.61 ± 0.05	0.21 ± 0.03	0.13 ± 0.02
10.	*Elaeocarpus tectorius* (Kernel)	0.78 ± 0.04	0.38 ± 0.03	0.49 ± 0.07

TABLE 5.24 *(Continued)*

Sl. No.	Name of the Plant	Components		
		Total Free Phenolic g 100g^{-1}	**Tannin g 100g^{-1}**	**Hydrogen Cyanide mg 100g^{-1}**
11.	*Eleusine coracana* (Seed)	0.16 ± 0.02	0.11 ± 0.01	0.08 ± 0.01
12.	*Ensete superbum* (Seed)	0.18 ± 0.02	0.11 ± 0.01	0.08 ± 0.01
13.	*Givotia rottleriformis* (Kernel)	0.28 ± 0.03	0.13 ± 0.01	0.40 ± 0.04
14.	*Heracleum rigens* var. *rigens*	0.32 ± 0.04	0.16 ± 0.03	0.14 ± 0.02
15.	*Impatiens balsamina* (Seed)	0.29 ± 0.05	0.11 ± 0.02	0.11 ± 0.01
16.	*Moringa concanensis* (Kernel)	1.31 ± 0.08	0.28 ± 0.03	0.26 ± 0.05
17.	*Ocimum gratissimum* (Seed)	0.28 ± 0.01	0.17 ± 0.02	0.20 ± 0.03
18.	*Oryza meyeriana* var. *granulata* (Grain)	0.09 ± 0.01	0.05 ± 0.01	0.06 ± 0.01
19.	*Pithecellobium dulce* (Aril)	0.15 ± 0.02	0.09 ± 0.02	0.11 ± 0.02
20.	*Sapindus emarginatus* (Seed)	0.34 ± 0.05	0.18 ± 0.03	0.09 ± 0.01
21.	*Sesamum indicum* (Seed)	0.36 ± 0.02	0.17 ± 0.01	0.17 ± 0.03
22.	*Shorea roxburghii* (Cotyledon)	0.42 ± 0.03	0.15 ± 0.02	0.14 ± 0.02
23.	*Sterculia foetida* (Kernel)	0.28 ± 0.01	0.15 ± 0.02	0.45 ± 0.03
24.	*Sterculia guttata* (Kernel)	0.18 ± 0.01	0.14 ± 0.01	0.18 ± 0.02
25.	*Sterculia urens* (Kernel)	0.28 ± 0.02	0.11 ± 0.01	0.10 ± 0.01
26.	*Strychnos nux-vomica* (Kernel)	0.13 ± 0.03	0.10 ± 0.01	0.26 ± 0.04
27.	*Terminalia bellirica* (Kernel)	0.23 ± 0.02	0.16 ± 0.02	0.24 ± 0.03
28.	*Terminalia chebula* (Kernel)	0.38 ± 0.03	0.18 ± 0.01	0.16 ± 0.02
29.	*Ziziphus rugosa* (Kernel)	0.24 ± 0.05	0.21 ± 0.03	0.25 ± 0.02
30.	*Ziziphus xylopyrus* (Kernel)	0.30 ± 0.04	0.17 ± 0.03	0.24 ± 0.03

[a] All the values are means of triplicate determinations expressed on dry weight basis.

± Denotes standard error.

It is a matter of great pride that among the 18 hot spots known for rich flora in the world over, two are located in India. They are the Eastern Himalayas and the Western Ghats (Khoshoo, 1996). It is to be recorded here that the hill chain of Western Ghats recognized as a region of the high level of biodiversity is under the threat of rapid loss of genetic resources (Gadgil, 1996). The bio-diverse nature of the Eastern Ghats is meager. Keeping these facts intact in mind, in the present study, the wild edible food plants have been surveyed and documented from the natural strands

TABLE 5.25 Antinutritional Factors of Underutilized Legumes[a]

Sl. No.	Name of the plant	Total Free Phenolic g 100 g^{-1}	Tannin g 100 g^{-1}	L-DOPA g 100 g^{-1}	Hydrogen Cyanide mg 100 g^{-1}
1.	*Atylosia scarabaeoides* (Seed)	0.28 ± 0.08	0.54 ± 0.01	2.28 ± 0.01	0.14 ± 0.07
2.	*Canavalia gladiata* (Seed)	0.09	0.84 ± 0.06	2.76 ± 0.25	0.71 ± 0.04
3.	*Canavalia virosa* (Seed)	1.86 ± 0.10	0.56 ± 0.08	1.24 ± 0.21	0.56 ± 0.03
4.	*Chamaecrista absus* (Seed)	0.89 ± 0.03	0.28 ± 0.04	0.56 ± 0.05	0.26 ± 0.02
5.	*Dolichos lablab* var. *vulgaris* (seed)	0.78 ± 0.06	0.38 ± 0.04	1.24 ± 0.08	0.24 ± 0.03
6.	*Dolichos trilobus* (Seed)	1.34 ± 0.09	0.58 ± 0.07	1.74 ± 0.11	0.33 ± 0.04
7.	*Entada rheedi* (Kernel)	3.13 ± 0.05	0.63 ± 0.04	2.23 ± 0.16	0.20 ± 0.05
8.	*Lablab purpureus* var. *lignosus* (Seed)	0.26 ± 0.03	0.40 ± 0.02	1.43 ± 0.11	0.46 ± 0.06
9.	*Lablab purpureus* var. *purpureus* (Seed)	0.38 ± 0.04	0.26 ± 0.01	0.98 ± 0.03	0.32 ± 0.06
10.	*Macrotyloma uniflorum* (seed)	0.98 ± 0.03	0.42 ± 0.03	1.10 ± 0.07	0.21 ± 0.05
11.	*Mucuna atropurpurea* (Kernel)	2.46 ± 0.15	0.03 ± 0.01	1.09 ± 0.07	0.38 ± 0.07
12.	*Mucuna pruriens* var. *pruriens* (Seed)	4.63 ± 0.28	0.24 ± 0.01	7.26 ± 0.09	0.26 ± 0.02
13.	*Mucuna pruriens* var. utilis (Black coloured seed coat) (Seed)	4.16 ± 0.18	0.28 ± 0.01	6.28 ± 0.26	0.18 ± 0.02
14.	*Mucuna pruriens* var. utilis (White coloured seed coat) (Seed)	3.82 ± 0.16	0.30 ± 0.02	7.56 ± 0.19	0.20 ± 0.03
15.	*Neonotonia wightii*. var. *coimbatorensis*. (Seed)	1.03 ± 0.08	0.12 ± 0.01	0.88 ± 0.06	0.18 ± 0.01
16.	*Rhynchosia cana* (Seed)	1.60 ± 0.12	0.44 ± 0.03	1.09 ± 0.07	0.27 ± 0.03
17.	*Rhynchosia filipes* (Seed)	0.17 ± 0.01	0.40 ± 0.05	1.82 ± 0.08	0.22 ± 0.04
18.	*Rhynchosia rufescens* (Seed)	1.88 ± 0.03	0.56 ± 0.06	0.88 ± 0.04	0.30 ± 0.01
19.	*Rhynchosia suaveolens* (Seed)	1.84 ± 0.10	0.68 ± 0.06	1.24 ± 0.02	0.25 ± 0.02
20.	*Tamarindus indica* (Kernel)	3.76 ± 0.14	0.03 ± 0.01	4.12 ± 0.36	0.09 ± 0.01
21.	*Teramnus labialis* (Seed)	2.01 ± 0.07	0.21 ± 0.02	1.52 ± 0.03	0.31 ± 0.03
22.	*Vigna bourneae* (Seed)	1.04 ± 0.03	0.36 ± 0.02	1.02 ± 0.03	0.16 ± 0.01
23.	*Vigna radiata* var. *sublobata* (Seed)	1.38 ± 0.03	0.79 ± 0.03	0.62 ± 0.02	0.36 ± 0.02
24.	*Vigna trilobata* (Seed)	1.27 ± 0.13	0.33 ± 0.02	1.48 ± 0.04	0.24 ± 0.01
25.	*Vigna unguiculata* subsp. *cylindrica* (Seed)	2.77 ± 0.12	0.77 ± 0.09	1.96 ± 0.05	0.59 ± 0.03
26.	*Vigna unguiculata* subsp. *unguiculata* (Seed)	2.08 ± 0.08	0.38 ± 0.02	0.73 ± 0.01	0.29 ± 0.02
27.	*Xylia xylocarpa* (Kernel)				

[a] All the values are means of triplicate determinations expressed on dry weight basis.
± Denotes standard error.

of the Southeastern slopes of the Western Ghats, Tamil Nadu, India. The Palliyars, the dominant tribal group, inhabit the locality of the study area. The present study focuses on the dependence of the Palliyars on edible plants and attempts at an exhaustive analysis of the nutritional qualities of such edible plants.

Plate - IV

Edible Tuber, Rhizome and Corm and Root Types

a. *Abelmoschus moschatus* Medik.

b. *Argyreia pilosa* Arn.

c. *Asparagus racemosus* Willd.

d. *Colocasia esculenta* (L.) Schott

e. *Curculigo orchioides* Gaertn.

f. *Curcuma neilgherrensis* Wt.

g. *Dolichos trilobus* L.

(See color insert.)

Plate - V

Edible Tubers, and Bulbils in their Natural Habitat

a. *Dioscorea oppositifolia* L. var. *dukhumensis*. Prian & Burkill

b. *Dioscorea oppositifolia* L. var. *oppositifolia*

c. *Dioscorea pentaphylla* L. var. *pentaphylla*

d. *Dioscorea pentaphylla* L. var. *pentaphylla* - Bulbils

e. *Dioscorea tomentosa* Koen. ex Spreng.

(See color insert.)

Plate - VI

Edible Tubers in their Natural Habitat

a. *Dioscorea oppositifolia* L. var. *dukhumensis*. Prian & Burkill

b. *Dioscorea oppositifolia* L. var. *oppositifolia*

c. *Dioscorea pentaphylla* L. var. *pentaphylla*

d. A tribal man holding a tuber of *Dioscorea tomentosa* Koen.ex Spreng

(See color insert.)

Plate - VII

Edible Rhizome and Corm in their Natural Habitat

a. *Canna indica* L.

b. *Maranta arundinacea* L.

c. *Xanthosoma violaceum* Schott.

d. *Xanthosoma sagittifolium* (L.) Schott.

(See color insert.)

Plate - VIII

Edible Tuber and Corm in their Natural Habitat

a. *Dioscorea esculenta* (Lour.) Burk.

b. *Manihot esculenta* Crantz

c. *Dioscorea wallichii* Hook.

d. *Alocasia macrorrhiza* Schott.

(See color insert.)

Plate - IX

Greens in their Natural Habitat

a. *Basella alba* L. var. *alba*

b. *Cardiospermum halicacabum* L.

c. *Celosia argentea* L.

d. *Commelina ensifolia* R.Br.

e. *Heracleum rigens* Wall. ex DC. var. *rigens Clarke*

f. *Jasminum calophyllum* Wall. ex A. DC.

(See color insert.)

Plate - X

Edible Flowers in their Natural Habitat

a. *Ipomoea alba* L. - Flower

b. *Ipomoea alba* L. with swollen special

c. *Moringa cancanensis* Nimmo ex Gibs.

d. *Phoenix pusilla* Gaertn.

Edible Stem and Apical Meristem

e. *Caralluma lasiantha* (Wight) N.E. Br

f. *Phoenix pusilla* Gaertn.

(See color insert.)

Plate - XI

Edible Unripe Fruits and Pods in their Natural Habitat

a. *Carissa carandas* L.

b. *Commiphora caudata* (Wight & Arn.) Engler

c. *Lablab purpureus (L.)* Sweet var. *lignosus* (Prain) Kumari comb.

d. *Senna occidentalis* (L.) Link.

e. *Solanum pubescens* Willd.

f. *Solanum torvum* Sw.

(See color insert.)

Plate - XII

Edible Ripe Fruits

a. *Carissa carandas* L.

b. *Ficus racemosa* L.

c. *Grewia heterotricha* Mast.

d. *Grewia hirsuta* Vahl.

e. *Lantana camara* L. var. *aculeata* (L.) Moid.

f. *Solanum torvum* Sw.

(See color insert.)

Plate - XIII

Edible Ripe Fruits

a. *Passiflora foetida* L.

b. *Phyllanthus emblica* L.

c. *Polyalthia cerasoides* (Roxb.) Bedd.

d. *Syzygium cumini* (L.) Vahl.

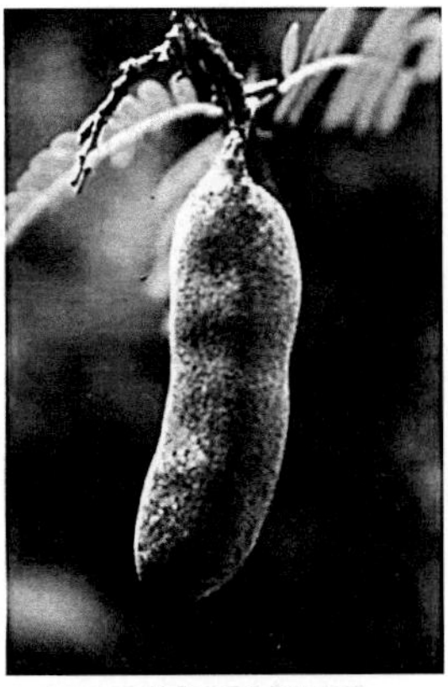

e. *Tamarindus indica* L.

f. *Uvaria rufa* Blume.

(See color insert.)

Plate - XIV

Edible Seeds in their Natural Habitat

a. *Atylosia scarabaeoides* (L.) Benth.

b. *Canavalia gladiata* (Jacq.) DC.

c. *Givotia rottleriformis* Griff.

d. *Oryza meyeriana (Zoll. & Mor.) Baill. var. granulata (Nees & Arn. ex Watt) Duist.*

e. *Rhynchosia filipes* Benth.

f. *Sterculia guttata* Roxb. ex DC.

(See color insert.)

Plate - XV

Edible Seeds

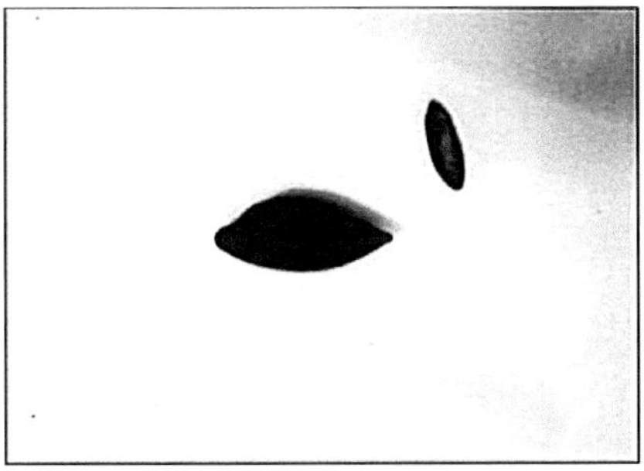

a. *Canarium strictum* Roxb.

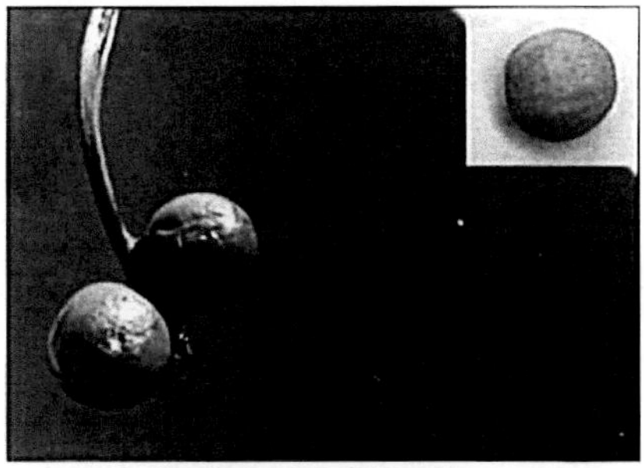

b. *Cycas circinalis* L.

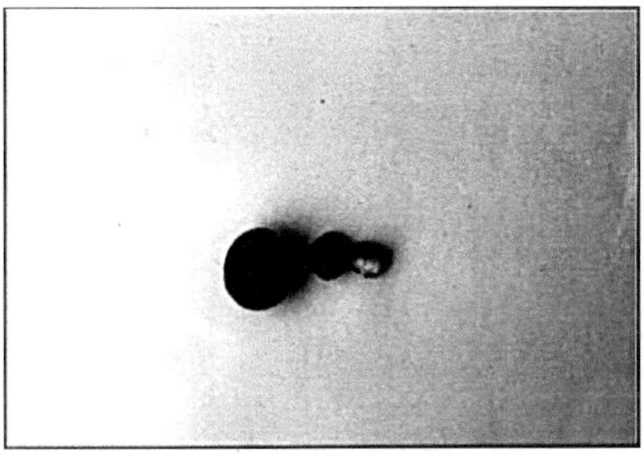

c. *Givotia rottleriformis* Griff.

d. *Sterculia foetida* L.

e. *Sterculia guttata* Roxb. ex DC.

f. *Terminalia bellirica* (Gaertn) Roxb.

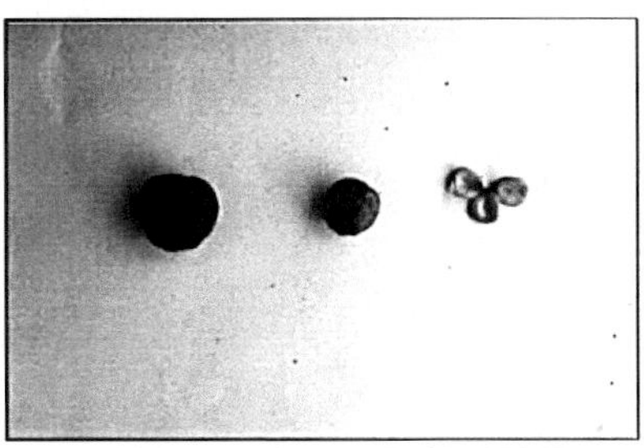

g. *Ziziphus xylopyrus* (Retz.) Willd.

(See color insert.)

Plate - XVI

Edible Seeds

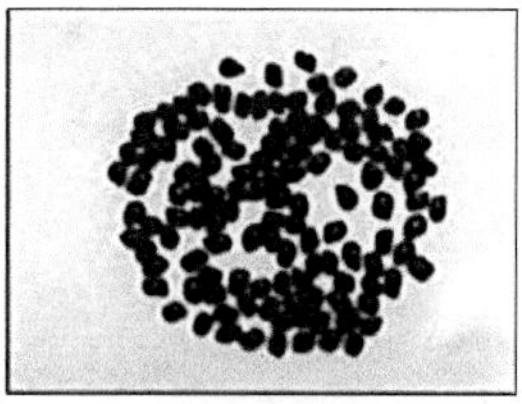

a. *Chamaecrista absus* (L.) Irwin & Barneby

b. *Entada rheedi* Spreng.

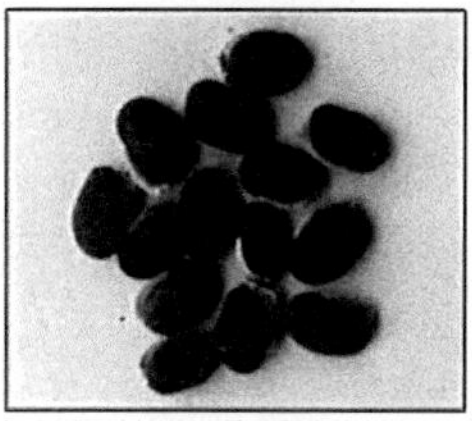

c. *Lablab purpureus (L.)* Sweet var. *lignosus* (Prain) Kumari comb.

d. *Mucuna atropurpurea* DC.

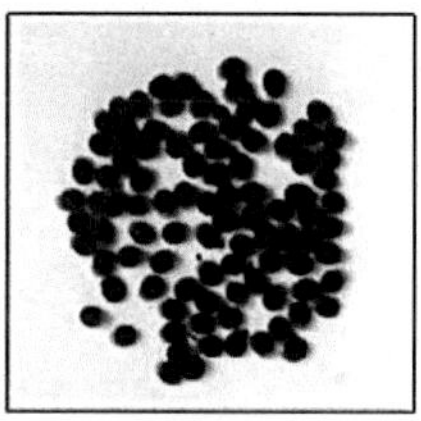

e. *Rhynchosia cana* DC.

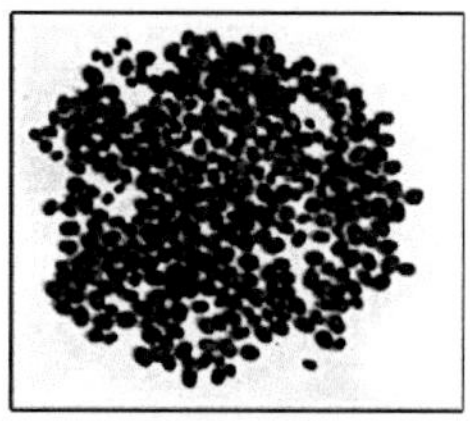

f. *Rhynchosia filipes* Benth.

g. *Teramnus labialis* (L.f) Spr.

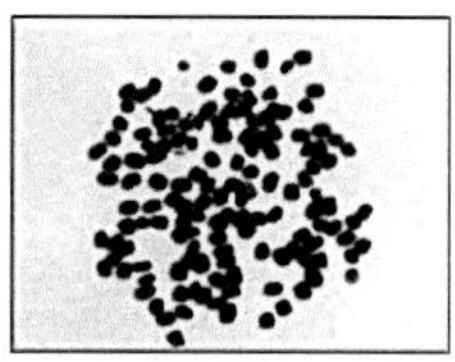

h. *Vigna trilobata* (L.) Verdc.

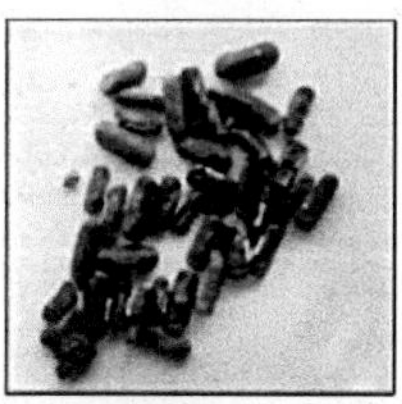

i. *Vigna unguiculata* (L.) Walp. subsp. *cylindrica* (L.) Eselt.

j. *Vigna unguiculata* (L.) Walp. subsp. *unguiculata*

(See color insert.)

Plate - I
Area Map

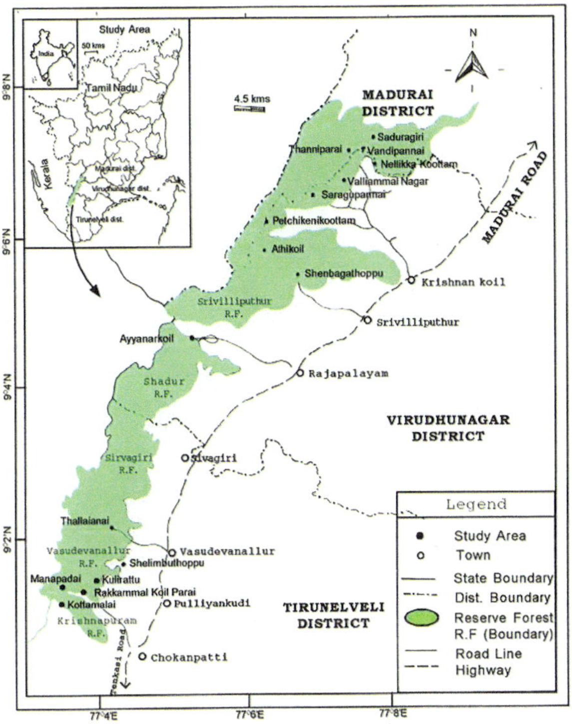

a. Grizzled Giant Squirrel Wildlife Sanctuary

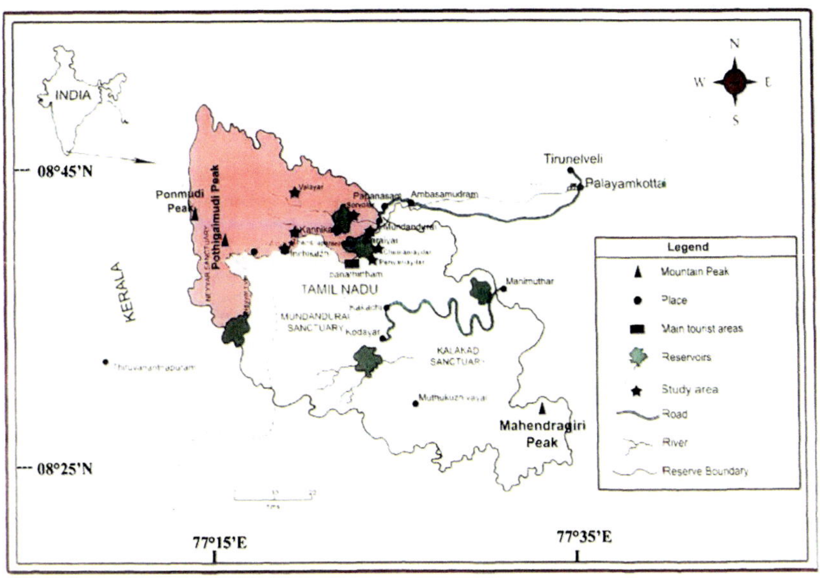

b. Agasthiarmalai Biosphere Reserve

Plate - II
Area Map

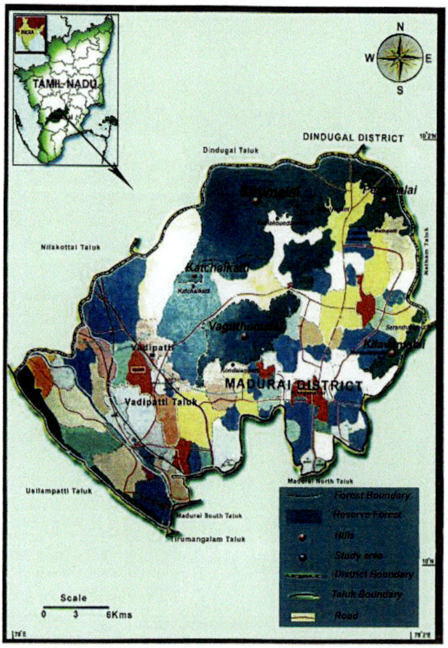

c. Madurai District

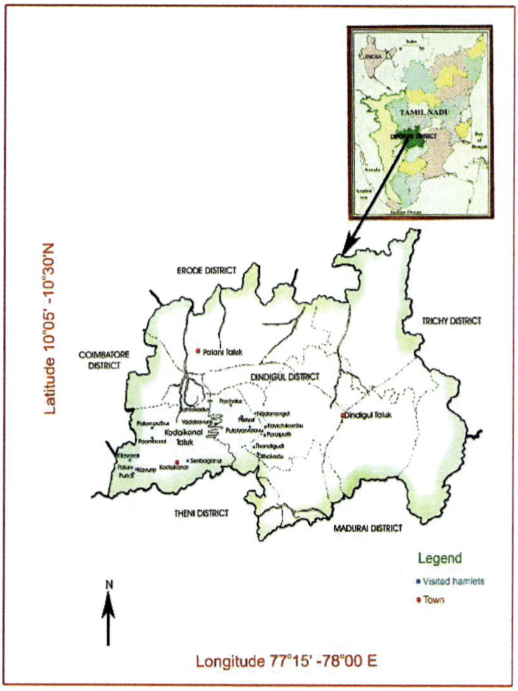

d. Palani Hills

Plate - III

Tribals of the Area Surveyed

a. An old Palliyar tribal man of Athikoil Sector of Grizzled Giant Squirrel Wildlife Sanctuary

b. A Palliyar family of Rakkammal Koil Parai

c. A very old Kani Woman

d. A Kanikkar family of Servalar

e. Chithirai, a Valaiyan holding *Ceropegia juncea* Roxb.

f. A Valaiyan's family of Valaiyapatti

g. A Pulayan Woman in Pachalur

h. A Pulayan family in Pachalur

Plate - IV

Edible Tuber, Rhizome and Corm and Root Types

a. *Abelmoschus moschatus* Medik.

b. *Argyreia pilosa* Arn.

c. *Asparagus racemosus* Willd.

d. *Colocasia esculenta* (L.) Schott

e. *Curculigo orchioides* Gaertn.

f. *Curcuma neilgherrensis* Wt.

g. *Dolichos trilobus* L.

Plate - V

Edible Tubers, and Bulbils in their Natural Habitat

a. *Dioscorea oppositifolia* L. var. *dukhumensis*. Prian & Burkill

b. *Dioscorea oppositifolia* L. var. *oppositifolia*

c. *Dioscorea pentaphylla* L. var. *pentaphylla*

d. *Dioscorea pentaphylla* L. var. *pentaphylla* - Bulbils

e. *Dioscorea tomentosa* Koen. ex Spreng.

Plate - VI

Edible Tubers in their Natural Habitat

a. *Dioscorea oppositifolia* L. var. *dukhumensis.* Prian & Burkill

b. *Dioscorea oppositifolia* L. var. *oppositifolia*

c. *Dioscorea pentaphylla* L. var. *pentaphylla*

d. A tribal man holding a tuber of *Dioscorea tomentosa* Koen.ex Spreng

Plate - VII

Edible Rhizome and Corm in their Natural Habitat

a. *Canna indica* L.

b. *Maranta arundinacea* L.

c. *Xanthosoma violaceum* Schott.

d. *Xanthosoma sagittifolium* (L.) Schott.

Plate - VIII

Edible Tuber and Corm in their Natural Habitat

a. *Dioscorea esculenta* (Lour.) Burk.

b. *Manihot esculenta* Crantz

c. *Dioscorea wallichii* Hook.

d. *Alocasia macrorrhiza* Schott.

Plate - IX

Greens in their Natural Habitat

a. *Basella alba* L. var. *alba*

b. *Cardiospermum halicacabum* L.

c. *Celosia argentea* L.

d. *Commelina ensifolia* R.Br.

e. *Heracleum rigens* Wall. ex DC. var. *rigens Clarke*

f. *Jasminum calophyllum* Wall. ex A. DC.

Plate - X

Edible Flowers in their Natural Habitat

a. *Ipomoea alba* L. - Flower

b. *Ipomoea alba* L. with swollen special

c. *Moringa cancanensis* Nimmo ex Gibs.

d. *Phoenix pusilla* Gaertn.

Edible Stem and Apical Meristem

e. *Caralluma lasiantha* (Wight) N.E. Br

f. *Phoenix pusilla* Gaertn.

Plate - XI

Edible Unripe Fruits and Pods in their Natural Habitat

a. *Carissa carandas* L.

b. *Commiphora caudata* (Wight & Arn.) Engler

c. *Lablab purpureus (L.)* Sweet var. *lignosus* (Prain) Kumari comb.

d. *Senna occidentalis* (L.) Link.

e. *Solanum pubescens* Willd.

f. *Solanum torvum* Sw.

Plate - XII

Edible Ripe Fruits

a. *Carissa carandas* L.

b. *Ficus racemosa* L.

c. *Grewia heterotricha* Mast.

d. *Grewia hirsuta* Vahl.

e. *Lantana camara* L. var. *aculeata* (L.) Moid.

f. *Solanum torvum* Sw.

Plate - XIII

Edible Ripe Fruits

a. *Passiflora foetida* L.

b. *Phyllanthus emblica* L.

c. *Polyalthia cerasoides* (Roxb.) Bedd.

d. *Syzygium cumini* (L.) Vahl.

e. *Tamarindus indica* L.

f. *Uvaria rufa* Blume.

Plate - XIV

Edible Seeds in their Natural Habitat

a. *Atylosia scarabaeoides* (L.) Benth.

b. *Canavalia gladiata* (Jacq.) DC.

c. *Givotia rottleriformis* Griff.

d. *Oryza meyeriana (Zoll. & Mor.) Baill. var. granulata (Nees & Arn. ex Watt) Duist.*

e. *Rhynchosia filipes* Benth.

f. *Sterculia guttata* Roxb. ex DC.

Plate - XV

Edible Seeds

a. *Canarium strictum* Roxb.

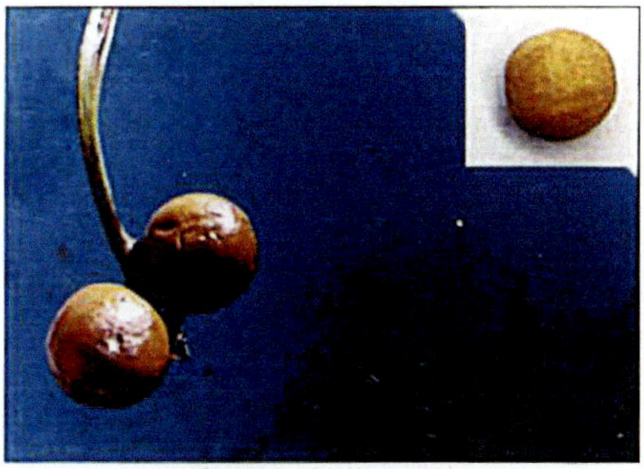

b. *Cycas circinalis* L.

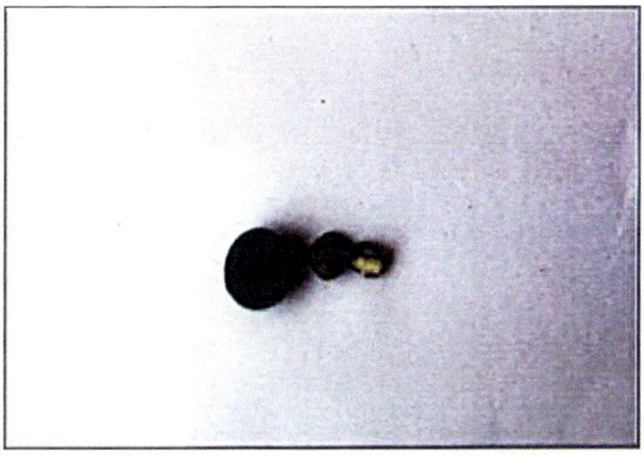

c. *Givotia rottleriformis* Griff.

d. *Sterculia foetida* L.

e. *Sterculia guttata* Roxb. ex DC.

f. *Terminalia bellirica* (Gaertn) Roxb.

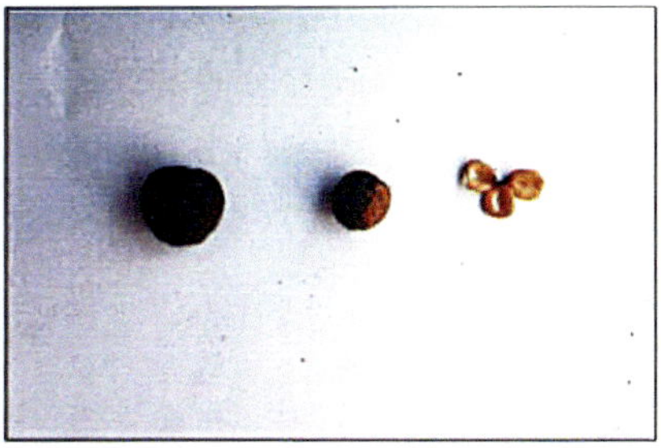

g. *Ziziphus xylopyrus* (Retz.) Willd.

Plate - XVI

Edible Seeds

a. *Chamaecrista absus* (L.) Irwin & Barneby

b. *Entada rheedi* Spreng.

c. *Lablab purpureus (L.)* Sweet var. *lignosus* (Prain) Kumari comb.

d. *Mucuna atropurpurea* DC.

e. *Rhynchosia cana* DC.

f. *Rhynchosia filipes* Benth.

g. *Teramnus labialis* (L.f) Spr.

h. *Vigna trilobata* (L.) Verdc.

i. *Vigna unguiculata* (L.) Walp. subsp. *cylindrica* (L.) Eselt.

j. *Vigna unguiculata* (L.) Walp. subsp. *unguiculata*

CHAPTER 6

Nutraceutical Analysis

The term 'nutraceutical' was introduced in 1989 by Stephen DeFelice. It can be defined as, a food (or part of a food) that provides medical or health benefits, including the prevention and/or treatment of a disease (Brower, 1998). In simple terms, nutraceuticals are those foods or parts of foods that afford health and/or medical advantages including prevention, protection, and healing of a disease (Belem, 1999). In view, of its medicinal synergy, economical status and no side effects, the nutraceuticals, functional or health foods have gained a wide interest during the last few decades (Raskin et al., 2002). The health-promoting effects of phytochemicals and nutraceuticals and/or functional foods are likely due to a complex mix of biochemical and cellular interactions which together promote the overall health of the individual (Dillard and German, 2000). The clinical success of nutraceutical products (Acharya and Thomas, 2007) coupled with increasing health consciousness results in the rapid global growth of the nutraceuticals and functional food industry (Hasler, 2000). The major chemical compounds recognized as potential health-promoting advantages are the phenolics, flavonoids, alkaloids, carotenoids, prebiotics, phytosterols, tannins, fatty acids, terpenoids, saponins, and soluble and insoluble dietary fibers (Patwardhan et al., 2005).

Nutraceutical is the hybrid of 'nutrition' and 'pharmaceutical.' Nutraceuticals are found in a mixture of products emerging from (a) the food industry, (b) the herbal and dietary complement market, (c) pharmaceutical industry, and (d) the newly merged pharmaceutical or agribusiness or nutrition conglomerates. It may range from isolated nutrients, herbal products, dietary supplements and diets to genetically engineered 'designer' foods and processed, manufactured goods such as cereals, soups, and beverages (Malik, 2008; Dureja et al., 2003).

Nutraceuticals enclose most of the therapeutic areas such as anti-arthritic, cold and cough, sleeping disorders, digestion and preclusion of certain cancers, osteoporosis, blood pressure, cholesterol control, pain killers, depression and diabetes (Pandey et al., 2000; Sami Labs, 2002).

As said by Rishi (2006) and Hathcock (2001), the nutraceutical industry include three main segments namely herbal/ natural products, dietary supplements, and functional foods. Among them, those most rapidly growing segments are the herbal/natural products and the dietary supplements (Nutrition Business Journal, 2006). In 2007, the world nutraceutical market grew to reach $74.7 billion, compared to that of 2002, when it reached $46.7 billion (BCC Research). The leading countries having nutraceutical markets include the USA, UK, and Japan (BCC Research).

Research and development are at the peak in this emerging nutraceutical field. The greatest scientific need pertains to standardization of the nutraceutical compounds or products carefully develop and perform clinical studies to provide the basis for health claims to produce an impact on the consumers as well as on the nutraceutical companies.

6.1 CATEGORIZING NUTRACEUTICALS

Nutraceuticals can be organized in several ways depending upon its easier understanding and application, e.g., for academic instruction, clinical trial design, functional food development or dietary recommendations. Some of the most common ways of classifying nutraceuticals can be based on food sources, mechanism of action, chemical nature, etc. The food sources used as nutraceuticals are all natural and can be categorized as (Kalia, 2005; Kokate et al., 2002): dietary fiber, probiotics, prebiotics, poly-unsaturated fatty acids, antioxidant vitamins, polyphenols, and spices.

In the next part of the analysis, a brief explanation of the health and medical benefits of some nutraceuticals are done. More broadly, nutraceuticals can be classified into two groups (Pandey et al., 2000):

i) potential nutraceuticals; and
ii) established nutraceuticals.

A potential nutraceutical could become an established one only after efficient clinical data of its health and medical benefits are obtained. It is to be noted that much of the nutraceutical products are still in the 'potential' category.

6.2 ROOT AND TUBER CROPS

Starchy root and tuber crops are second only in importance to cereals as global sources of carbohydrates. They supply a substantial part of the world's food supply and also an important source of animal feed and processed the item for consumptions for human beings and industrial use. Starchy roots and tubers are plants which store edible starch material in subterranean stems, roots, rhizomes, corms, and tubers. They are derived from various botanical sources.

Plants producing starchy roots, tubers, rhizomes, corms, and stems are imperative for nutrition and health. They play an indispensable role in the diet of populations in developing countries in addition to their practice for animal feed and for manufacturing starch, alcohol and fermented foods and beverages. Nutritionally, roots and tubers have a great potential to offer economical sources of dietary energy, in the variety of carbohydrates. The energy from tubers is about one-third of that of an equivalent weight of rice or wheat due to high dampness of tubers. However, high yields of roots and tubers give more energy per land unit per day in contrast to cereal grains (FAO, 1990).

In general, the protein content of roots and tubers is low ranging from 1 to 2% on a dry weight basis (FAO, 1990). Potatoes and yams contain high amounts of proteins among other tubers. Sulfur-containing amino acids, namely methionine, and cysteine, are the limiting ones in root crop proteins. Cassava, sweet potatoes, potatoes, and yam contain a few vitamin C and yellow varieties of sweet potatoes, yam and cassava contain β-carotene. Taro is a superior source of potassium. Roots and tubers are deficient in most other vitamins and minerals but contain momentous amounts of dietary fiber (FAO, 1990). Similar to other crops, the nutritional value of roots and tubers varies with variety, location, soil type, and agricultural practices. Tubers and root crops are major sources of a number of compounds namely, saponins, phenolic compounds, glycoalkaloids, phytic acids, carotenoids, and ascorbic acid. Several bioactivities, namely antioxidant, immune-modulatory, antimicrobial, antidiabetic, antiobesity, and hypercholesterolemic activities are reported for tubers and root crops.

Since the acceptance of the convention on biodiversity in 1992, there has been a general harmony on the importance of biodiversity, particularly the diversity of wild and cultivated plants, to fill the need of the world population for food (FAO, 1998). In developing countries like India, people do not get enough food to meet their daily requirement, and most often the diet is deficient in one or more micronutrients (FAO, 1994). India faced a series of famines and major food shortages before 1940s. National food grain production was merely 50.82 million tons during 1950–1951, but has risen to 264.38 million tons in 2012–2013 (FAO, 1994). Edible roots and tubers enrich the diet due to the presence of starch and energy supplemented metabolites in them. They also possess medicinal properties due to the presence of diverse secondary metabolites.

The part of roots and tubers to the energy supply in different populations varies with the country. The relative importance of these crops is evident through their annual global production which is approximately 836 million tonnes (FAOSTAT, 2013). Asia is the main producer followed by Africa, Europe, and America. Asian and African regions manufactured 43 and 33% correspondingly, of the global production of roots and tubers. A number of species and varieties are consumed, but cassava, potatoes, and sweet potatoes include 90% global production of root and tuber crops (FAOSTAT, 2013).

6.2.1 CASSAVA (MANIHOT ESCULENTA)

Cassava is mainly cultivated root crop in the tropics. It is to be noted here that, because of the long growth season (8–24 months), its production is limited to the tropical and subtropical regions in the world. Cassava is a perennial shrub fitting into the family Euphorbiaceae. The genus *Manihot* comprises 98 species. *M. esculenta* is the most extensively cultivated member (Nassar et al., 2008). Cassava instigated in South America and was later distributed to tropical and subtropical regions of Africa and Asia (Blagbrough et al., 2010). Cassava plays an imperative role as a staple for more than 500 million people in the world due to its high carbohydrate content (Blagbrough et al., 2010). A number of bioactive compounds, namely, cyanogenic glucosides such as linamarin and lotaustralin, noncyanogenic glucosides, hydroxycoumarins such as scopoletin, terpenoids, and flavonoids, are accounted in cassava roots (Blagbrough et al., 2010; Prawat et al., 1995; Reilly et al., 2004).

6.2.2 AROIDS

Aroids are tuber or underground stem bearing plants fitting in the family Araceae. There are quite a lot of edible tubers/stems such as taro (*Colocasia*), giant taro (*Alocasia*), tannia or yautia (*Xanthosoma*), elephant foot yam (*Amorphophallus*), and swamp taro (*Cyrtosperma*). The source of tannia is South America and the Caribbean regions (FAO, 1999). *Colocasia*, originating in India and Southeast Asia, is a staple food in many islands of the South Pacific, such as Tonga and Western Samoa, and in Papua New Guinea. Furthermore, taro is the most widely cultivated crop in Asia, Africa, and Pacific as well as the Caribbean Islands.

6.3 MINOR TUBER CROPS

6.3.1 CANNA

Canna is rhizomatous type tuber which is broadly spread throughout the tropics and subtropics. The genus *Canna* belongs to the family Cannaceae. The edible types of *Canna edulis* instigated in the Andean region or Peruvian coast and extended from Venezuela to northern Chile, in South America. It is commercially cultivated in Australia for the production of starch.

6.3.2 ARROWROOT

Maranta arundinacea L. (West Indian arrowroot) is cultivated for its edible rhizomes. It belongs to *Marantaceae* and is believed to have originated in the Northwestern part of South America. Arrowroot has been widely distributed throughout the tropical countries like India, Sri Lanka, Indonesia, the Philippines, and Australia and the West Indies.

6.3.3 DIOSCOREA SPECIES

The genus *Dioscorea* belongs to the family Dioscoreaceae. This family is the most well-known within the order Dioscoreales (Burkill, 1960; Ayensu and Coursey, 1972; Dansi et al., 1999; Tamiru et al., 2008; APG III, 2009). The family is supposed to be one among the earliest angiosperms, and it is most likely to be originated in Southeast Asia (Coursey, 1967). The various *Dioscorea* species apparently followed a divergent evolution in three continents separated by the formation of the Atlantic Ocean and desiccation of the Middle East (Hahn, 1995). Therefore, these major species occur in three isolated centers: West Africa, Southeast Asia, and Tropical America (Alexander and Coursey, 1969). These centers are also considered areas for independent yam domestication and represent considerable diversity (Asiedu et al., 1997).

The tuber crop under study, *Dioscorea*, is superior to many others as an essential medico food used by about 300 million people throughout the world (Arnau et al., 2010). In fact, they are one of the principal sources of energy food for many people in the Tropics (Nayaboga et al., 2014). As per source of dietary nutrients, *Dioscorea* species rank as the world's fourth most important root and tuber crops after potatoes, cassava, and sweet potatoes (Lev and Shriver, 1998). Many of the tubers of *Dioscorea* are bitter in taste. The local people living there use traditional skills to remove this bitterness. Aboriginals also make use of their tubers as snacks, and in roasted, powdered, and other forms (Kumar et al., 2012; Misra et al., 2013). The same processes are also followed in various parts of India, including the Himalayan regions and North-Eastern part of India (Sheikh et al., 2013), Orissa (Sinha and Lakra, 2005; Kumar and Satpathy, 2011; Kumar et al., 2012; Misra et al., 2013), Tamil Nadu (Rajyalakshmi and Geervani, 1994; Shajeela et al., 2011), and among Palliyar and Kanikkar tribes (Shanthakumari et al., 2008; Arinathan et al., 2009) living in South-eastern slopes of Western Ghats (Padmaja et al., 2001). They are also used by the local people of Kumaon and Garhwal hills of India (Pramila et al., 1991). The nutrient content of the Yam has been evaluated with several other crops (Wanasundera and Ravindran, 1994).

Out of about 600 species, only 10 species of *Dioscorea* are grown throughout the world. In India, about 26 *Dioscorea* species are reported. Out of them, 13 are reported in SBR, Odisha (Kumar et al., 2012). Further, than 13 species, only one, *D. alata* L., is cultivated in this region, and the leftover species mostly grow wild in this zone. Tubers of 12 wild *Dioscorea* species available in Odisha have sound nutritional values, but have low sweetness due to their bitterness (Kumar et al., 2012). *Dioscorea* the starchy edible tuber is of ample economic and nutritional consequence in the Tropical and Sub-tropical regions of the world (Sharma and Bastakoti, 2009). Indeed, the tubers are rich sources of food and energy. The world's estimated yam production in 2010 was 48.7million tons (Abasi et al., 2013). They have been reported to be good sources of essential dietary nutrients (Wanasundera and Ravindran, 1994).

6.3.3.1 ETHNOBOTANICAL VALUES OF DIOSCOREA SP.

Dioscorea have sound ethnobotanical values throughout the Tropics. There are numerous reports available on local claims on *Dioscorea* species worldwide. In some forest areas of Southern Thailand, which are situated in the Tropical rain forest zone of Southeast Asia, local people use *Dioscorea* species to treat warts (Maneenoon et al., 2008). The boiled tubers of *D. membranacea* are used to treat asthma and fever (Maneenoon et al., 2008). The mucilage from the tubers of *D. piscatorum* is used to poison fish, and it is also used by the native people of Malaysia as a piscicide (Burkill, 1951, 1960). *D. prazeri* is used as soap and shampoo to kill lice in India (Maneenoon et al., 2008). *Dioscorea* is used in healing gastritis among Yoruba local groups of Cuba (Kadiri et al., 2014). Tubers of *D. hamiltonii* are replaced as body refrigerant during summer and are also employed to treat diarrhea (Dutta, 2015). *D. bulbifera* is used against tuberculosis and raw tuber of *D. pentaphylla* against diphtheria in cattle (Sharma and Bastakoti, 2009). Tubers of *D. oppositifolia* are used in the healing of swellings, scorpion stings, and snake bites (Dutta, 2015). Juice of *D. wallichii* is used to cure Jaundice. *D. hispida* is used as an antidote to arrow poison (Sinha and Lakra, 2005; Edison et al., 2006; Mishra et al., 2008; Swarnkar and Katewa, 2008; Sahu et al., 2010). Details of ethnobotanical values of different species of *Dioscorea* are listed in Table 6.1.

6.4 BIOACTIVITIES OF PHYTOCHEMICALS IN ROOTS AND TUBERS

6.4.1 ANTIOXIDANT ACTIVITY

Accumulating research evidences demonstrate that oxidative stress plays a major role in the development of several chronic diseases such as different types of cancer, cardiovascular diseases, arthritis, diabetes, autoimmune and neurodegenerative disorders, and aging. Though internal antioxidant defense systems, either enzymes (superoxide dismutase, catalase, and glutathione peroxidase) or other compounds (lipoic acid, uric acid, ascorbic acid, tocopherol, and glutathione) are available in the body, external sources of antioxidants are needed, as internal defense system may get overwhelmed by excessive exposure to oxidative stress. A number of researches have reported the antioxidant activities of a number of roots and tuber crops.

Water yam (*Dioscorea alata*) was accounted to possess the highest DPPH radical scavenging activity of 96% among diverse selected tuber

TABLE 6.1 Ethnobotanical Values of Common Dioscorea Species of SBR, India

Botanical Name	Plant Parts	Ethno-botanical/Pharmacological Values	Supporting Literature
Dioscorea alata L.	Tubers	Tuber powder is used to cure piles	Jadhav et al., 2011
		Tubers are eaten raw twice a day until weakness is reduced	Kamble et al., 2010
		Juice of tuber is used to kill stomach worm	Samanta and Biswas, 2009
Dioscorea bulbifera L.	Leaves	Paste is used against skin infections	Girach et al., 1999
	Stem	Tender shoots and twigs are crushed and rubbed on wet hair to remove dandruff	Dutta, 2015
	Tubers	Raw tuber is eaten to enhance appetite	Mishra, R.K. et al., 2008
		Tuber is good for intestinal coli, relieving dysmenorrhea, reducing acidity against, against rheumatoid arthritis, to relieve intense inflammation in the acute phrase, in spasmodic asthma, for menopausal problems, for labor pain and the prevention of early miscarriage, for hernia, relieving the pain of childbirth.	Nayak et al., 2004; Patil and Patil, 2005; Bhogaonkar and Kadam, 2006; Mehta and Bhatt, 2007
		Bulbils are used to reduce throat pain	Mbiantcha et al., 2011
		Boiled tubers are taken orally to reduce body heat	Singh et al., 2009
		Used against boils and dysentery	Nag, 1999
		Tuber powder mix with butter is given to check diarrhea	Jadhav et al., 2011
		Used as refrigerant to reduce body heat during summer	Dutta, 2015
		Used to treat skin infection	Tiwari and Pande, 2006
		Used to treat bronchial cough and used as antiseptic	Bhatt and Negi, 2006
		Useful for acidity and ulcers	Dutta, 2015
		Root paste mixed with cow milk is taken orally for the treatment of cough and asthma	Teron, 2011
		Used to treat typhoid with *Curcuma aromatica*	Jain et al., 2008
		Used in ulcer, piles, syphilis and dysentery and powder and used to kill hair lice	Abhyankar and Upadhyay, 2011

TABLE 6.1 *(Continued)*

Botanical Name	Plant Parts	Ethno-botanical/Pharmacological Values	Supporting Literature
		About 10 gm of powder is given once a day for 5-6 days after menses as contraceptive	Kamble et al., 2010
		Tubers are boiled after processing and given for abdominal pains. The tubers are dried and pea sized pieces are cut and given in early morning with water for 3 days to cure piles	Abhyankar and Upadhyay, 2011
		The roasted and mashed tubers are eaten with salt to cure cough	Singh et al., 2009
Dioscorea dumetorum (Kunth) Pax	Tuber	Tuber juice is used to make arrow poison	Edison et al., 2006
		Tuber paste is used to relieve pain and to treat boils, dysentery and swellings	
Dioscorea esculenta (Lour.) Burkil	Tuber	Tubers are sued for treatment of chest pain, nervous disorders and swellings	Edison et al., 2006
		Tuber paste is used to relieve pain and to treat boils, dysentery and swellings	Dutta et al., 2015
Dioscorea hispida Dennst.	Tuber	Water of soaked tuber is used as medicine for eyes	Meena and Yadav, 2011
		Used as fish poison	Nashriyah et al., 2011
		Sap of tuber is pasted around the affected parts and covered with cloths for about one night to treat peeling of skin of feet.	Sharma and Bastakoti, 2009
		Tuber is used to treat vomiting, indigestion, possesses narcotic properties and fresh tuber taken as purgative	Dutta, 2015
		Tubers are roasted and pounded and its paste is applied on wounds and injuries	Kamble et al., 2010

TABLE 6.1 *(Continued)*

Botanical Name	Plant Parts	Ethno-botanical/Pharmacological Values	Supporting Literature
Dioscorea oppositifolia L.	Leaf	Leaf paste is used as antiseptic for ulcers	Sheikh et al., 2013
	Tuber	Tuber is boiled with *D. uniflorus* and is given to women once a day for nearly a month after delivery to revive their strength	Mishra et al., 2008
		Oral administration of tuber powder mixed with honey is used for increasing sperm.	Sharma and Bastakoti, 2009
		Powdered root mixed with cow urine is applied on scorpion bite	Nashriya et al., 2011
		Leaves are mixed with leaves of Clematis and 2-3 drops of juice put in the nose of affected person to get relief after sneezing in fits and epilepsy	Kamble et al., 2010
Dioscorea pentaphylla L.	Tuber	Tubers are applied on swelling of joints and used as tonic to improve body immunity	Edison et al., 2006
		Used for stomach pain	Choudhary et al., 2008
		Crushed mass of tuber is given to cattle when they become sick by eating green leaves of maize	Sharma and Bastakoti, 2009
		Tuber is used as tonic and also to cure stomach troubles and rheumatic swellings	Dutta, 2015
		Inflorescence is used as vegetables for body weakness	Kamble et al., 2010
		Tubers are useful to allay pain and swelling	
Dioscorea wallichii Hook. f.	Tuber	Roasted and eaten for flatulence	Dutta, 2015
		Used in stomach pain	Rout and Panda, 2015

crops such as sweet potato, potato, cocoyam, and other *Dioscorea* yams (Dilworth et al., 2012).

Cornago et al. (2011) showed the Tubers Phenolic Content (TPC) and antioxidant activities of two major Philippine yams of *Ube* (purple yam) and *Tugui* (lesser yam). Purple yam (*Dioscorea alata*) and lesser yam (*Dioscorea esculenta*) had a TPC which ranged from 69.9 to 421.8 mg GAE/100 g dry weight. The purple yam variety *Daking* had the highest TPC and antioxidant activities as determined by DPPH radical scavenging activity, reducing power and ferrous ion chelating capacity, whereas *Sampero* and *Kimabajo* showed the lowest TPC and antioxidant activities.

Hsu et al. (2011) studied the antioxidant activity of water and ethanolic extracts of yam peel on tert-butyl hydroperoxide (t-BHP). The result showed an induced oxidative stress in mouse liver cells (Hepa 1–6 and FL83B). Ethanolic extracts of yam peel exhibited a better protective effect on t-BHP treated cells compared to that of water extracts. Besides, it was observed that ethanolic extract increased catalase activity, whereas water extract decreased it. In accordance with Chen and Lin (2007) heating affected the TPC, antioxidant capacity, and the stability of dioscorin of various yam tubers. Raw yams contained higher TPC than their cooked matching parts. Moreover, the DPPH radical scavenging activities declined with increasing temperature. TPC and dioscorin content of yam cultivars (*Dioscorea alata* var. Tainung number 2) and keelung yam (*D. japonica* var. *pseudojaponica)* showed a relationship with DPPH radical scavenging activity and ferrous ion chelating effect. Phytochemicals of yams seem to increase the activities of endogenous antioxidant enzymes. The administration of yams decreased the levels of glutamyl transpeptidase (GGT), low-density lipoprotein, and triacylglycerol in the serum of rats in which hepatic fibrosis was persuaded by carbon tetrachloride. Treatment of rats with yams increased the antioxidant activities of hepatic enzymes, that is, glutathione peroxidase and superoxide dismutase (Chan et al., 2010).

A few researches have reported the antioxidant activities of cassava roots. A recent study (Omar et al., 2012) showed that the antioxidant activities of organically grown cassava tubers were higher properties than those of mineral-base fertilized roots. They established that TPC and flavonoid content (FC) were appreciably higher for organic cassava compared to those of cassava grown using inorganic fertilizers.

6.4.2 ANTICANCER ACTIVITIES

Cancer is a principal cause of death worldwide, and it is mostly communicated through unhealthy food habits and lifestyle. It is important to find ways to reduce and prevent the risk of cancer through dietary components, which are present in plant foods. Cancer is a multistage disease condition and tapping at any initial stage could help soothe the disease condition. Root and tuber phytochemicals have proved anticancer effects in several types of carcinoma cell lines and animal models.

In addition to phenolic compounds, saponins present in roots and tubers play a fundamental role as anticancer/anti-tumor agents. There are numerous groups of saponins, namely, cycloar-tanes, ammaranes, oleananes, lupanes, and steroids, demonstrating a strong anti-tumor effect on many types of cancers. For instance, cycloartane saponins possess anti-tumor properties in human colon cancer cells and tumor xenografts. They down-regulated expression of the hepatocellular carcinoma (HCC) tumor marker fetoprotein and suppressed HepG2 cell growth by inducing apoptosis and modulating an extracellular signal-regulated protein kinase-(ERK-) independent NF- B signaling pathway. Besides, oleananes saponins exerted their anti-tumor effect through various pathways, such as anticancer, antimetastasis, immunostimulation, and chemoprevention pathways (Auyeung et al., 2009).

Several reports have shown that glycoalkaloids such as alpha-chaconine and alpha-solanine seen in tubers are potential anticarcinogenic agents. Glycoalkaloids showed antiproliferative activities against human colon (HT-29) and liver (HepG2) cancer cells as assessed by the MTT assay (Lee et al., 2004).

Wang et al. (2011) accounted that the aqueous extract of yam (*Dioscorea alata*) inhibited the $H_2O_{2-}CuSO_4$ induced smash up of calf thymus DNA and protected human lymphoblastoid cells from $CuSO_4$ induced DNA damage. Extract of water yam contains a homogenous compound with a single copper-binding site and also is a good natural, safe (redox inactive) copper chelator. In addition to phenolic compounds, saponins and mucilage polysaccharides present in yams are responsible for this activity. Additionally, water-soluble mucilage polysaccharides are the most important copper chelators in the extract of water yam. Thus *Dioscorea alata* aqueous extracts could serve as potential agents in the organization of copper-mediated oxidative disorders and diabetes.

6.4.3 IMMUNOMODULATORY ACTIVITIES

Purified dioscorin from yam tubers proved immune-modulatory activities in vitro. The effects of dioscorin on native BALB/c mice spleen cell proliferation were assessed by MTT assay. It was found that dioscorin stimulated RAW 264.7 cells to produce nitric oxide (NO), in the absence of lipopolysaccharide (LPS) pollution. Yam dioscorin exhibited immune-modulatory activities by the innate immunity which is a nonspecific immune system. This comprises the cells and mechanisms that defend the host from infection by other organisms in a nonspecific manner. Dioscorin was reported to stimulate cytokine production and to enhance phagocytosis. Additionally, the released cytokines may act synergistically with phytohemagglutinin (PHA) which is a lectin found in plants that stimulate the proliferation of splenocytes (Liu et al., 2007).

Quite a lot of readings have demonstrated the immune activity of yam mucopolysaccharides (YMP). *In vitro* cytotoxic activity of mouse splenocyte against leukemia cell was increased in the presence of YMP of *Dioscorea batatas* at 10 g/mL. Also the production of IFN- was radically increased in the YMP treated splenocytes, suggesting their capability of inducing cell-mediated immune responses. In addition, YMP at a concentration of 50 g/mL boosted up the uptaking capacity and lysosomal phosphatase activity of peritoneal macrophages (Choi et al., 2004).

Dioscorea phytocompounds developed murine splenocyte proliferation *ex vivo* and advanced regeneration of bone marrow cells in vivo. Mice which were fed with a *Dioscorea* extract recovered damaged bone marrow progenitor cell populations that had been exhausted by large doses of 5-fluorouracil (5-FU). Moreover, they reported that the phytocompound(s) responsible for these bioactivities had a high molecular weight (≥100 kDa) and were most likely polysaccharides. They postulated those high-molecular-weight polysaccharides in DsCE-II act on specific target cell types in the GI tract (dendritic cells, intestinal epithelial cells, and T-cells) to arbitrate a cascade of immunoregulatory activities leading to the recovery of damaged cell populations following 5-FU or other chemical insults in the bone marrow, spleen, or other immune cell systems (Su et al., 2011).

Dioscorea tuber mucilage from Taiwanese yams (*Dioscorea japonica)* showed significant effects on the innate immunity and adaptive immunity on BALB/c mice through oral administration. As well, it was found that the specific antibodies rapidly responded against foreign proteins (or antigens) in the presence of yam mucilage. Mucilage from these yam varieties showed signs of a stimulatory effect on phagocytic activity by granulocyte and monocyte (*ex vivo*), on peritoneal macrophages, and on the RAW 264.7 cells (*in vivo*) of mice (Shang et al., 2007).

Yams (*Dioscorea esculenta*) showed anti-inflammatory activity on carrageenan-induced edema in the right hind paw of Wistar rats. However, this activity was short lived as it was quickly removed from the system after reaching the peak within 2 hours. Phytochemical screening of *D. esculenta* substantiated the presence of saponins, sitosterol, stigmasterol, cardiac glycosides, fats, starch, and diosgenin, which could be responsible for the observed activity (Olayemi and Ajaiyeoba, 2007). Diosgenin contained in Chinese yam was an immunoactive steroidal saponin which also showed a prebiotic effect. Diosgenin had also beneficial effects on the growth of enteric lactic acid bacteria (Chen et al., 2003).

Methanolic extracts of *A. campanulatus* (Suran) tuber also showed immunomodulatory activity. Authors indicated that the presence of steroids and flavonoids in *A. campanulatus* (Suran) tuber may be responsible for the observed immunomodulatory activity (Tripathi et al., 2010).

6.4.4 ANTIMICROBIAL ACTIVITY

Yam varieties with their phenolic compounds are potential agents with antimicrobial efficacy. Sonibare and Abegunde (2012) reported that the methanolic extracts of *Dioscorea* yams (*Dioscorea dumetorum* and *Dioscorea hirtiflora*) showed antioxidant and antimicrobial activities. Antimicrobial activity was determined by the agar diffusion method (for bacteria) and pour plate method (for fungi). Non-edible *D. dumetorum* showed the highest in vitro antibacterial activity against *Proteus mirabilis*. The methanolic extracts from *D. hirtiflora* made obvious antimicrobial activity against all tested organisms, specifically, *Staphylococcus aureus, E. coli, Bacillus subtilis, Proteus mirabilis, Salmonella typhi, Candida albicans, Aspergillus niger*, and *Penicillium chrysogenum*.

6.4.5 HYPOGLYCEMIC ACTIVITIES

Diabetes mellitus is a chronic disorder noticeable by elevated levels of glucose in the blood and life-threatening complications that can, unfortunately, lead to death.

Ethanolic extract of tubers of *Dioscorea alata* showed an antidiabetic activity in alloxan-induced diabetic rats. Diabetic rats with administered yam extract exhibited significantly lower creatinine levels which could be a result of an improved renal function by reduced plasma glucose level and subsequent glycosylation of renal basement membranes. Several bioactive compounds, counting phenolics, were identified in the ethanolic extract of *D. alata*. These consist of hydro-Q9 chromene, tocopherol, feruloyl glycerol, dioscorin, cyanidin-3-glucoside, catechin, procyani-din, cyanidin, peonidin 3-gentiobioside, and alatanins A, B, and C (Maithili et al., 2011).

6.4.6 HYPOCHOLESTEROLEMIC ACTIVITY

Cardiovascular diseases are among the leading causes of death worldwide. It is well known that diet plays an important role in the regulation of cholesterol homeostasis. External agents possessing anticholesterolemic activities continuously show beneficial effects on risk reduction and management of the disease conditions.

Diosgenin, a steroidal saponin of yam (*Dioscorea*), demonstrated antioxidative and hypolipidemic effects in vivo. Rats nourished with a high-cholesterol diet were supplemented with either 0.1 or 0.5% diosgenin for 6 weeks. The lipid profile of the plasma and liver, lipid peroxidation and antioxidant enzyme activities in the plasma, erythrocyte, and gene expression of anti-oxidative enzymes in the liver and the oxidative DNA damage in lymphocytes were determined. Diosgenin showed pancreatic lipase inhibitory activity, the protective effect of liver under high-cholesterol diet, reduced total cholesterol level, and protection in opposition to the oxidative damaging effects of polyunsaturated fatty acids (Son et al., 2007). Steroidal saponins of yams are exercised for industrial drug processing. Saponins, such as dioscin and gracilin, and prosapogenins of dioscin have long been famous from yam. The content of dioscin was about 2.7% (w/w). Diosgenin content was about 0.004 and 0.12–0.48% in cultivated yam and wild yam, correspondingly. The antihypercholesterolemic effect of yam saponin is related to its inhibitory activity against cholesterol absorption (Ma et al., 2002).

The effect of yam diosgenin on hypercholesterolemia had been also reported by Cayen and Dvornik (1979). Hypercholesterolemic rats fed with yam (*Dioscorea*) showed that diosgenin had little cholesterol absorption, increased hepatic cholesterol production, and increased biliary cholesterol discharge without affecting serum cholesterol level. In concurrence with this finding, several studies showed that diosgenin, in some *Dioscorea*, could improve fecal bile acid secretion and decrease intestinal cholesterol absorption (Thewles et al., 1993; Uchida et al., 1984). The relative contribution of biliary secretion and intestinal absorption of cholesterol in diosgenin stimulated fecal cholesterol excretion were studied using wild-type (WT) and Niemann-Pick C1-Like 1 (NPC1L1) knockout (LIKO) mice. NPC1L1 was in recent times identified as an indispensable protein for intestinal cholesterol absorption (Zhi-Dong et al., 2009). Diosgenin extensively increased biliary cholesterol and hepatic expression of cholesterol synthetic genes in both WT and LIKO mice. In addition, diosgenin stimulation of fecal cholesterol excretion was primarily attributable to its impact on hepatic cholesterol metabolism rather than NPC1L1-dependent intestinal cholesterol absorption (Temel et al., 2009).

Protein of dioscorin decontaminated from *D. alata* (cv. Tainong number 1) (TN1-dioscorin) and its peptic hydrolysates presented ACE inhibitor activities in a dose-dependent manner (Hsu et al., 2002). According to kinetic analysis, dioscorin showed mixed non-competitive inhibition against ACE. Dioscorin from *Dioscorea* might be beneficial in controlling high blood pressure (Temel et al., 2009).

Chen et al. (2003) reported the effects of Taiwan's yam (*Dioscorea alata* cv. Tainung number 2) on mucosal hydrolase activities and lipid metabolism in male Balb/c mice. High level of Tainung number 2 yam in the diet (50% w/w) reduced plasma and hepatic cholesterol levels and increased fecal steroid excretions in mice model. This could be due to the loss of bile acid in the enterohepatic cycle to fecal excretion. They further suggested that the increased viscosity of the digest and the thickness of the unstirred layer in the small intestine caused by Tainung number 2 yam fiber (or/and mucilage) decreased the absorption of fat, cholesterol, and bile acid. Short term (3-week) consumption of 25% Tainung number 2 yam in the diet could reduce the atherogenic index but not total cholesterol level in nonhypercholesterolemic mice. Also, additional dietary yam (50% yam diet) could consistently exert hypocholesterolemic effects in these mice. However, Diosgenin was not elucidated in Tainung number 2 used in this study. Thus, the authors suggested that diosgenin might not be involved in the cholesterol-lowering effect of Tainung number 2 yam. Dietary fiber and viscous mucilage could be active components for the beneficial cholesterol-lowering effects of yam. Besides, short term consumption (3-week) of 25% uncooked Keelung yam effectively reduced total blood cholesterol levels and the atherogenic index in mice. Authors pointed out that the active components for the lipid-lowering effects may be characteristic to dietary fiber, mucilage, plant sterols, or synergism of these active components.

6.4.7 HORMONAL ACTIVITIES

Yam (*Dioscorea*) has the capability to reduce the risk of cancer and cardiovascular diseases in postmenopausal women. It was shown that the levels of serum estrogen and sex hormone binding globulin (SHBG) increased significantly after subjects had been on a yam diet for 30 days. Furthermore, three serum hormone parameters measured, namely, estrogen, estradiol, and SHBG, did not change in those who were fed with sweet potatoes as the control. The risk of breast cancer which increased by estrogens might be balanced by the elevated SHB and the ratio of estrogen plus estradiol to SHBG. Authors further showed that high SHBG levels had a protective effect against the occurrence of type 2 diabetes mellitus and coronary heart diseases in women (Wu et al., 2005).

Chronic administration of *Dioscorea* may enhance bone strength and provide insight into the role of *Dioscorea* in bone remodeling and osteoporosis during the menopause. Administration of *Dioscorea* to ovariectomized rats brought down the porosity effect on bones and amplified the ultimate force of bones. The changes in biochemical and physiological functions seen in these animals were akin to those in menopausal women (Chen et al., 2008).

6.5 GRAIN LEGUMES FOR FOOD

In terms of economic importance, the Leguminosae are the most important family in the Dicotyledonae (Harborne, 1994). The Leguminosae is one of the largest families of the flowering plants with 18,000 species classified into around 650 genera (Polhill and Raven, 1981). This is just under a twelfth of all known flowering plants. The Leguminosae constitute one of the most important groups of plants used by the human race. In providing food crops for world agriculture legumes are second only to the grasses (cereals). In comparison to cereal grains, the seeds of legumes are rich in high-quality protein, providing man with a highly nutritious food resource. The chief staple foods such as beans, soybean, lentils, peas, and chickpeas, are all legumes. The total world value for leguminous crops is considered to be approximately two billion US dollars per annum. Many supplementary legumes are local food plants.

6.6 BENEFICIAL EFFECTS OF LEGUME SEEDS

Grain legumes can scarcely be found in the old books of traditional medicine as specific therapeutic agents. Some information can be found concerning minor and local species, especially in the eastern and far eastern countries (India and China). Other indications of phytotherapeutic use of legumes come from the Mediterranean regions, where grain legumes have traditionally been consumed for centuries and constitute an element of the Mediterranean diet (DeFeo et al., 1992; Guarrera, 2005). On the other hand, several suggestions considering the role of pulses in the prevention of relevant diseases, typical of the affluent countries, are available and represent the basis for the healthy claims on legume seeds.

As reminded by Kushi et al., 1999 the observation that diets low in meat and high in cereals and legumes are beneficial for health was noted at least as far back as the Old Testament in the first chapter of the Book of Daniel. More recent, but still vague, reports on pulse beneficial effects on man health and acknowledged contributing agents (rarely proteins) are summarized in the following subsections.

6.6.1.1 CARDIOVASCULAR DISEASE (CVD)

The frequent intake of dry legumes, in parallel with a saturated-fat poor diet, can help to control the lipid homeostasis and consequently reduce the risk of CVD. The legume high fiber content, their low glycaemic index and the presence of minor components, such as phytosterols, saponins, oligosaccharides, etc., are considered the main responsible agents for this property.

6.6.1.2 DIABETES

Dry legumes are claimed to help glycaemic control in diabetic individuals because of the low glycaemic index (GI) and the high content of undigestible fibers. Moreover, dry legumes may contribute to prevent insulin resistance, which represents the prodrome to type II diabetes.

6.6.1.3 DIGESTIVE TRACT DISEASES

The accelerated transit of the digested food in the intestinal tract and its final excretion play a number of positive effects, decreased re-absorption of cholesterol, incomplete starch digestion, lowering of fermentation processes. These factors have been considered beneficial in the prevention of cancer, especially colon cancer. Legumes are thought to play an important role in this respect.

6.6.1.4 OVERWEIGHT AND OBESITY

Despite their content of lipids, starch, and proteins, dry legumes are claimed to help maintaining a regular body weight, thanks to their great satiety effect, thus limiting the overall food intake daily. Various seed components have been claimed to bring about this effect.

6.7 HORSE GRAM

Horse gram (*Macrotyloma uniflorum* Lam. (Verdc.)) is an underutilized pulse crop grown in a wide range of adverse climatic conditions. It occupies an important place in human nutrition and has a rich source of protein, minerals, and vitamins. Besides nutritional importance, it has been linked to reduced risk of various diseases due to the presence of non-nutritive bioactive substances. These bioactive substances such as phytic acid, phenolic acid, fiber, enzymatic/proteinase inhibitors have significant metabolic and/or physiological effects. The significance of horse gram was well accepted by the folk/alternative/traditional medicine as a potential therapeutic agent to treat kidney stones, urinary diseases, piles, common cold, throat infection, fever, etc. Due to the inception of nutraceutical concept and increasing health consciousness, the requirement of nutraceutical and functional food is increased. In recent years, isolation and utilization of potential antioxidants from legumes including horse gram are increased as it decreases the risk of intestinal diseases, diabetes, coronary heart disease, prevention of dental caries, etc.

6.7.1 ETHNIC USES OF HORSE GRAM

The soup extract from kulattha (Horse gram), called yusa, which was consumed commonly during the Sutra period (c. 1500–800 BC) are the rasams of today (Achaya, 1998). Horse gram is widely grown for human food as a pulse and fodder crop for livestock (Cook et al., 2005) as well as green manure and medicinal crop. In rural areas, seeds of horse gram are consumed after dehydration followed by boiling or frying (Purseglove, 1974) along with cooked rice, sorghum or pearl millet. Sprouted seeds, having high nutritional quality, are widely consumed by the native tribal people (Bravo et al., 1999). Even now, in addition to its nutritive value, the consumption of sprouted seed become increasingly popular due to their excellent source to reduce the risk of various diseases and exerting health-promoting effects (Pasko et al., 2009). In Indian traditional medicine, seeds of horse gram are used for treating urinary stones (Yadava and Vyas, 1994; Ravishankar and Vishnupriya, 2012), urinary diseases and piles (Yadava and Vyas, 1994), regulate the abnormal menstrual cycle in women (Neelam, 2007), act as astringent, tonic (Brink, 2006), and also used to treat calculus afflictions, corpulence, hiccups, and worms (Chunekar and Pandey, 1998). Furthermore, the cooked liquor of the horse gram seeds with spices is considered to be a potential remedy for the common cold, throat infection, fever, and the soup said to generate heat (Perumal and Sellamuthu, 2007).

Owing to their nutritional and medicinal value, there is an increased demand to explore an underutilized legume (Chel-Guerrero et al., 2002; Arinathan et al., 2003) to alleviate malnutrition and reduce the risk of various diseases in developing countries. Horse gram is an excellent source of protein (17.9–25.3%), carbohydrates (51.9–60.9%), essential amino acids, energy, low content of lipid (0.58–2.06%), iron (Bravo et al., 1999; Sodani et al., 2004), molybdenum (Bravo et al., 1999), phosphorus, iron and vitamins such as carotene, thiamine, riboflavin, niacin and vitamin C (Sodani et al., 2004).

6.8 MUCUNA SPECIES

The genus *Mucuna*, belonging to the Fabaceae family, sub-family Papilionaceae, consists of approximately 150 species of annual and perennial legumes. The itching bean, *Mucuna pruriens* (L.) DC var. *pruriens* is an underutilized legume species grown predominantly in Asia, Africa and in parts of America (Vadivel and Janardhanan 2000). Mature seeds, seeds from unripe pods and young pods of itching beans are soaked and boiled/roasted and eaten as such or mixed with salt by the north-east Indian tribes; Khasi, Naga, Kuki, Jaintia, Chakma and Mizo (Arora, 1991); northwestern part of Madhya Pradesh tribes; Abujh-Maria, Maria, Muria, Gond and Halba (Sahu, 1996); South Indian tribes; Mundari and Dravidian (Jain, 1981); Kani, Kader and Muthuvan (Radhakrishnan et al., 1996) and Savera Jatapu, Gadebe and Kondadora (Rajyalakshmi and Geervani, 1994).

The mature seeds of velvet bean, *Mucuna pruriens* (L.) DC var. *utilis* (Wall.ex Wight) Bak. ex Burck are consumed by a South Indian hill tribe, the Kanikkars, after repeated boiling (Janardhanan and Lakshmanan, 1985). Recently Dravidian tribes in the Tirunelveli district have started cultivating it for use as a pulse (Janardhanan et al., 2003). Various preparation of this bean is also traditionally consumed in several parts of Srilanka by low-income groups (Ravindran and Ravindran, 1988). In parts of Asia, and Africa, the seeds are roasted and eaten (Haq, 1983).

In India, the roasted kernels of *Mucuna atropurpurea* is known to be consumed by the Palliyar tribals living in Grizzled Giant Squirrel Wildlife Sanctuary, Srivilliputhur, Southeastern Slope of the Western Ghats, Tamil Nadu (Arinathan et al., 2007).

Most *Mucuna* spp. exhibit reasonable tolerance to a number of abiotic stresses, including drought, low soil fertility, and high soil acidity, although they are sensitive to frost and grow poorly in cold, wet soils (Duke, 1981). The genus thrives best under warm, moist conditions, below 1500 m above sea level, and in areas with abundant rainfall. Like most legumes, the velvet bean has the potential to fix atmospheric nitrogen via a symbiotic relationship with soil microorganisms.

Among the various underutilized wild legumes, the velvet bean *Mucuna pruriens* is widespread in tropical and sub-tropical regions of the world. It is considered a viable source of dietary proteins (Janardhanan et al., 2003; Pugalenthi et al., 2005) due to its high protein concentration (23–35%) in addition its digestibility, which is comparable to that of other pulses such as soybean, rice bean, and lima bean (Gurumoorthi et al., 2003). It is therefore regarded as a good source of food. *Mucuna pruriens* var. *utilis* (black and white colored seed coat) *Mucuna pruriens* var. *pruriens,* and *Mucuna deeringiana* are underutilized legume species having the reliability to be a rich protein source. This proximate composition, mineral profiles, vitamins (niacin and ascorbic acid), protein fractions, fatty acid profiles, amino acid profiles of total seed protein, in vitro protein digestibility and anti-nutritional potential of the above said *Mucuna* varieties/species were assessed. The major findings of the study were as follows: crude protein content ranged from 23.76–29.74%, crude lipid 8.24–9.26%, total dietary fiber, 6.54–8.76%, ash 4.78–5.30%, carbohydrates 56.56–63.88%, and calorific value 1783.02–1791.60kJ100g-^{1}DM. The examined seed samples contained minerals such as Na, K, Ca, Mg and P in abundance. The fatty acid profiles reveal that the seed lipids contained higher concentrations of linoleic and palmitic acid.

The concentration of the non-protein amino acid L-Dopa in *Mucuna pruriens* var. *pruriens* and *Mucuna pruriens* var. *utilis* (black and white colored seed coat) was high when compared with those of the other earlier reports in *Mucuna gigantea, Mucuna monosperma,* all the accessions of *Mucuna atropurpurea* and Ayyanarkoil and Anaikatti accessions of *Mucuna pruriens* var. *pruriens* (Rajaram and Janardhanan, 1991; Mohan and Janardhanan, 1995; Kala et al., 2010b; Kalidass and Mohan, 2011). The high range of L-Dopa is encouraging from the point of view of pharmaceutical industries. Cultivator difference (Jeyaweera, 1981) and accessions variations (Mary Josephine and Janardhanan, 1992) are known to exist in the L-Dopa content of *Mucuna* beans.

On the medicinal side, the non-protein amino acid, L-dopa, is extracted from the seeds of *Mucuna bean* to provide commercial drugs for the treatment of Parkinson's disease (Haq, 1983). The seed powder is well-known to exhibit faster hypothermic (Rajendran et al., 1996) and anti-Parkinsonian activity than synthetic L-dopa (Hussain and Manyam, 1997). The seed powder is known to kindle more sexual activity in male albino rats than L-dopa and also is reported to arouse sexual desire in patients suffering from Parkinson's disease (Ananthakumar et al., 1994). Vigorex-SF capsule, an ayurvedic herbo-mineral formulation consisting of the seed powder is found to have the adaptogenic effect to improve libido, disturbed due to psychological fear and emotional imbalance and other allied ailments (Shaw and Bera, 1993). Seeds of this wild legume are widely used for treating male sexual dysfunction in Unani medicine (Amin et al., 1996). The blocking effect of King Cobra venom at the neuromuscular junction is removed by the aqueous extract of the seed of this plant species (Aguiyi et al., 1997). Rhinax, an herbal formulation comprising this wild pulse possesses anti-hepatotoxic activity (Dhuley and Naik, 1997). The tribe, Garos of Meghalaya, India consumes the seeds for growing potency, and the hairs of the pod are employed as a vermifuge (Vasudeva Rao and Shanpru, 1991). In Nigeria, powdered hairs on pods are administrated with honey for expelling intestinal worms (Gill and Nyawuame, 1994).

Among the various under-utilized wild legumes, the velvet bean *Mucuna pruriens* is widespread in tropical and sub-tropical regions of the world. It is considered a viable source of dietary proteins (Janardhanan et al., 2003; Pugalenthi et al., 2005) due to its high protein concentration (23–35%) in addition its digestibility, which is comparable to that of other pulses such as soybean, rice bean, and lima bean (Gurumoorthi et al., 2003). It is therefore regarded as a good source of food.

The dozen or so cultivated *Mucuna* spp. found in the tropics probably result from fragmentation deriving from the Asian cultigen, and there are numerous crosses and hybrids (Bailey and Bailey, 1976). The main differences among cultivated species are in the characteristics of the pubescence on the pod, the seed color, and the number of days to harvest of the pod. ‘Cowitch’ and ‘cowhage’ are the common English names of *Mucuna* types with abundant, long stinging hairs on the pod. Human contact results in an intensely itchy dermatitis, caused by mucunain (Infante et al., 1990). The nonstinging types known as ‘velvet bean’ have apprised as silky hairs.

The plant *M. pruriens*, widely known as ‘velvet bean, ‘ is a vigorous annual climbing legume originally from southern China and eastern India, where it was at one time widely cultivated as a green vegetable crop (Duke, 1981). It is one of the most popular green crops currently known in the tropics; velvet beans have great prospective as both food and feed as suggested by experiences worldwide. The velvet bean has been traditionally used as a food source by certain ethnic groups in a number of countries. It is cultivated in Asia, America, Africa, and the Pacific Islands, where its pods are used as a vegetable for human utilization and its young leaves are used as animal fodder (Lampariello et al., 2012).

Mucuna spp. has been reported to contain the toxic compounds L-dopa and hallucinogenic tryptamines, and anti-nutritional factors such as phenols and tannins (Awang et al., 1997). Due to the high concentrations of L-dopa (4–7%), velvet bean is a commercial source of this substance, used in the treatment of Parkinson's disease. The toxicity of unprocessed velvet bean may explain why the plant exhibits low susceptibility to insect pests (Duke, 1981). Velvet bean is well known for its nematicidic effects. It also reportedly possesses notable allelopathic activity, which may function to suppress competing plants (Gliessman et al., 1981).

Despite its toxic properties, various species of *Mucuna* are grown as a trivial food crop. Raw velvet bean seeds contain roughly 27% protein and are rich in minerals (Duke, 1981). During the 18th and 19th centuries, *Mucuna* was grown widely as a green vegetable in the foothills and lower hills of the eastern Himalayas and in Mauritius. Both the green pods and the mature beans were boiled and eaten. In Guatemala and Mexico, *M. pruriens* has for at least several decades been roasted and ground to make a coffee substitute; the seeds are widely known in the region as 'Nescafe,' in recognition of this use.

6.8.1 MUCUNA PRURIENS AS TRADITIONAL MEDICINE

M. pruriens is a popular Indian medicinal plant, which has long been used in traditional Ayurvedic Indian medicine, for diseases including parkinsonism. This plant is widely used in Ayurveda, which is an ancient traditional medical science that has been practiced in India since the *Vedic* times (1500–1000 BC). *M. pruriens* is reported to contain L-dopa as one of its constituents (Chaudhri, 1996). The beans have also been employed as a powerful aphrodisiac in Ayurveda (Amin, 1996) and have been used to treat nervous disorders and arthritis (Jeyaweera, 1981). The bean, if applied as a paste on scorpion stings, is thought to absorb the poison (Jeyaweera, 1981).

The non-protein amino acid-derived L-dopa (3,4-dihydroxyphenylalanine) found in this underutilized legume seed resists attack from insects, and thus controls biological invasion during storage. According to D'Mello (1995), all anti-nutritional compounds confer insect and disease resistance to plants. Further, L-dopa has been extracted from the seeds to provide commercial drugs for the treatment of Parkinson's disease. L-Dopa is a potent neurotransmitter precursor that is believed, in part, to be responsible for the toxicity of the *Mucuna* seeds (Lorenzetti et al., 1998). Antiepileptic and antineoplastic activity of methanol extract of *M. pruriens* has been reported (Gupta et al., 1997). A methanol extract of *M. pruriens* seeds has demonstrated significant *in vitro* antioxidant activity, and there are also indications that methanol extracts of *M. pruriens* may be a potential source of natural antioxidants and antimicrobial agents (Rajeshwar et al., 2005).

All parts of *M. pruriens* possess valuable medicinal properties, and it has been investigated in various contexts, including for its antidiabetic, aphrodisiac, antineoplastic, antiepileptic, and antimicrobial activities (Sathiyanarayanan et al., 2007). Its antivenom activities have been investigated by Guerranti et al. (2002), and its antihelminthic activity has been demonstrated by Jalalpure (2007). *M. pruriens* has also been shown to be neuroprotective (Misra and Wagner, 2007), and has demonstrated analgesic and anti-inflammatory activity (Hishika et al., 1981).

6.8.2 PHARMACOLOGICAL EFFECTS OF MUCUNA PRURIENS

All parts of the *Mucuna* plant possess medicinal properties (Sathiyanarayanan and Arulmozhi, 2007). *In vitro* and *in vivo* studies on *M. pruriens* extracts have revealed the presence of substances that exhibit a wide variety of pharmacological effects, including anti-diabetic, anti-inflammatory, neuroprotective and antioxidant properties, probably due to the presence of L-dopa, a precursor of the neurotransmitter dopamine (Misra and Wagner, 2007). It is known that the main phenolic compound of *Mucuna* seeds is L-dopa (approximately 5%) (Vadivel and Pugalenthi, 2008). Nowadays, *Mucuna* is widely studied because L-dopa is a substance used as a first-line treatment for Parkinson's disease. Some studies indicate that L-dopa derived from *M. pruriens* has much compensation over synthetic L-dopa when administered to Parkinson's patients, as synthetic L-dopa can have numerous side effects when used for many years.

In small amounts (approximately 0.25%) L-dopa corresponds to methylated and non-methylated tetrahydroisoquinoline (Siddhuraju and Becker, 2001; Misra and Wagner, 2004). These substances are present in the *Mucuna* roots, stems, leaves, and seeds. Other substances are present in different parts of the plant, among which are *N,N*-dimethyl tryptamine and some indole compounds (Tripathi and Updhyay, 2001). Alcoholic extracts of the seeds were shown to have potential anti-oxidant activity in *in-vivo* models of lipid peroxidation induced by stress (Tripathi and

Updhyay, 2001). On the other hand, Spencer et al. (1996) have reported that the prooxidant and antioxidant actions of L-dopa and its metabolites promote oxidative DNA damage and could also be harmful to tissues damaged by neurodegenerative diseases, namely parkinsonism. Moreover, a study using *in vitro* models revealed that L-dopa significantly increases the levels of oxidized glutathione in rat brain striatal synaptosomes (Spina et al., 1988). The observed depletion of reduced glutathione (GSH) could be due to the generation of reactive semiquinones from L-dopa (Spencer et al., 1995).

6.8.3 PROTECTIVE EFFECT OF MUCUNA PRURIENS SEEDS AGAINST SNAKE VENOM POISONING

M. pruriens is one of the plants that have been shown to be active against snake venom and, indeed, its seeds are used in traditional medicine to prevent the toxic effects of snake bites, which are mainly triggered by potent toxins such as neurotoxins, cardiotoxins, cytotoxins, phospholipase A_2 (PLA_2), and proteases (Guerranti et al., 2002). In Plateau State, Nigeria, the seed is prescribed as a prophylactic oral anti-snakebite remedy by traditional practitioners, and it is claimed that when the seeds are swallowed intact, the individual is protected for one full year against the effects of any snake bite (Guerranti et al., 2001). The mechanisms of the protective effects exerted by *M. pruriens* seed aqueous extract (MPE), were explored in detail, in a study involving the effects of Echis carinatus venom (EV) (Guerranti et al., 2002). In vivo experiments on mice showed that protection against the poison is evident at 24 hours (short-term), and 1 month (long term) after injection of MPE (Guerranti et al., 2008). MPE protects mice against the toxic effects of EV via an immune mechanism (Guerranti et al., 2002). MPE contains an immunogenic component, a multiform glycoprotein, which stimulates the production of antibodies that cross-react with (bind to) certain venom proteins (Guerranti et al., 2004). This glycoprotein, called gpMuc is composed of seven different isoforms with molecular weights between 20.3 and 28.7 kDa, and pI between 4.8 and 6.5 (Di Patrizi et al., 2006).

6.8.4 *ANTIMICROBIAL PROPERTIES OF MUCUNA PRURIENS LEAVES*

Various parts of certain plants are known to contain substances that can be used for therapeutic purposes or as precursors for the production of useful drugs (Sofowora, 1982). Plant-based antimicrobials represent a vast untapped source of medicines, and further investigation of plant antimicrobials is needed. Anti-microbials of plant origin have massive therapeutic potential. Phytochemical compounds are reportedly responsible for the anti-microbial properties of certain plants (Mandal et al., 2005). While bioactive compounds are often extracted from whole plants, the concentration of such compounds within the different parts of the plant varies. Parts known to contain the highest concentration of the compounds are preferred for therapeutic purposes. Some of these active components operate individually, others in combination, to inhibit the life processes of microbes, particularly pathogens. Crude methanolic extracts of *M. pruriens* leaves have been shown to have mild activity against some bacteria in experimental settings, probably due to the presence of phenols and tannins (Ogundare and Olorunfemi, 2007). Further studies are required in order to isolate the bioactive components responsible for the observed anti-microbial activity.

6.8.5 *NEUROPROTECTIVE EFFECT OF MUCUNA PRURIENS SEEDS*

In India, the seeds of *M. pruriens* have traditionally been used as a nervine tonic, and as an aphrodisiac for male virility. The pods are anthelmintic, and the seeds are anti-inflammatory. Powdered seeds possess anti-parkinsonism properties, possibly due to the presence of L-dopa (a precursor of neurotransmitter dopamine). It is well known that dopamine is a neurotransmitter. The dopamine content in brain tissue is reduced when the conversion of tyrosine to L-dopa is blocked. L-Dopa, the precursor of dopamine, can cross the blood-brain barrier and undergo conversion to dopamine, restoring neurotransmission (Kulhalli, 1999). Good yields of L-dopa can be extracted from *M. pruriens* seeds with EtOH-H_2O (1:1), using ascorbic acid as a protector (Misra and Wagner, 2007). An n-propanol extract of *M. pruriens* seeds yields the highest response in neuroprotective testing involving the growth and survival of DA neurons in culture. Captivatingly, n-propanol extracts, which contain a small amount of L-dopa, have shown significant neuroprotective activity, suggesting that a whole extract of *M. pruriens* seeds could be superior to pure L-dopa with regard to the treatment of parkinsonism.

6.8.6 ANTIDIABETIC EFFECT OF MUCUNA PRURIENS SEEDS

Using a combination of chromatographic and NMR techniques, the presence of d-chiro-inositol and its two galacto-derivatives, O-α-d-galactopyranosyl-(1→2)-d-*chiro*-inositol (FP1) and O-α-d-galactopyranosyl-(1→6)-O-α-d-galactopyranosyl-(1→2)-D-*chiro*-inositol (FP2), was demonstrated in *M. pruriens* seeds (Donati et al., 2005). Galactopyranosyl d-chiro-inositols are relatively rare and have been isolated recently from the seeds of certain plants; they constitute a minor component of the sucrose fraction of *Glycine max* (Fabaceae) and lupins, and a major component of *Fagopyrum esculentum* (Polygonaceae) (Horbovitz et al., 1998). Although usually ignored in phytochemical analyses conducted for dietary purposes, the presence of these cyclitols is of interest due to the insulin-mimetic effect of d-chiro-inositol, which constitutes a novel signaling system for the control of glucose metabolism (Larner et al., 1998; Ortmeyer et al., 1995). According to Akhtar et al., (1990), *M. pruriens* seeds used at a dose of 500 mg/kg reduced plasma glucose levels. These and other data demonstrated that the amount of seeds necessary to obtain a significant anti-diabetic effect contain a total of approximately 7 mg of d-*chiro*-inositol (including both free, and that derived from the hydrolysis of FP1 and FP2). The anti-diabetic properties of *M. pruriens* seed EtOH/H_2O 1:1 extract are most likely due to d-*chiro*-inositol and its galacto-derivatives.

6.8.7 ANTIOXIDANT ACTIVITY OF MUCUNA PRURIENS

Free radicals that have one or more unpaired electrons are produced during normal and pathological cell metabolism. Reactive oxygen species (ROS) react readily with free radicals to become radicals themselves. Antioxidants provide protection to living organisms from damage caused by uncontrolled production of ROS and concomitant lipid peroxidation, protein damage and DNA strand breakage. Several substances from natural sources have been shown to contain antioxidants and are under study. Anti-oxidant compounds such as phenolic acids, polyphenols, and flavonoids, scavenge free radicals such as peroxide, hydroperoxide or lipid peroxyl, and thus inhibit oxidative mechanisms. Polyphenols are important phytochemicals due to their free radical scavenging and *in vivo* biological activities (Bravo et al., 1998); the total polyphenolic content has been tested using Folin-Ciocalteau reagent. Flavonoids are simple phenolic compounds that have been reported to possess a wide spectrum of biochemical properties, including anti-oxidant, anti-mutagenic and anti-carcinogenic activity (Beta et al., 2005). The hydrogen donating ability of the methanol extract of *M. pruriens* was computed in the presence of 1,1-diphenyl-2-picrylhydrazyl (DPPH) radical. In a current study, Kumar et al. (2010) found that ethyl acetate and methanolic extract of whole *M. pruriens* plant (MEMP), which contains large amounts of phenolic compounds, exhibits high antioxidant and free radical scavenging activities. These *in vitro* assays indicate that this plant extract is a significant source of natural antioxidant, which may be useful in preventing various oxidative stresses. It has been reported (Ujowundu et al., 2010) that methanolic extracts of *M. pruriens* leaves have numerous biochemical and physiological activities, and contain pharmaceutically valuable compounds.

Mucuna pruriens is an exceptional plant. On the one hand, it is a good source of food, as it is rich in crude protein, essential fatty acids, starch content, and certain essential amino acids. On the other hand, it also contains various anti-nutritional factors, such as protease inhibitors, total phenolics, oligosaccharides (raffinose, stachyose, verbascose), and some cyclitols with anti-diabetic effects. In fact, all parts of the *Mucuna* plant possess medicinal properties. The main phenolic compound is L-dopa (5%), and *M. pruriens* seeds include some components that are able to inhibit snake venom. In addition, methanolic extracts of *M. pruriens* leaves have demonstrated anti-microbial and anti-oxidant activities in the presence of bioactive compounds such as phenols, polyphenols, and tannins, and preliminary studies on keratinocytes support its possible topical usage to treat redox-driven skin diseases. Collectively, the studies cited in this review suggest that this plant and its extracts may be of therapeutic value with regard to several pathologies, although further work is needed to investigate in more detail the mechanisms underlying the pharmacological activities of *M. pruriens*.

CHAPTER 7

Conclusion

Wild edible plants are worldwide distributed but in very low quantity. Due to their difficult geography and climatic conditions that may not be suitable for human survival, these plants are generally in the majority in places that are not affected by human interaction. Southeastern Slopes of Western Ghats is characterized by a rich diversity of plants as well as the rich heritage of wild edible plants system. The traditional knowledge about the use of indigenous edible plants has been explored; therefore, the ethnological knowledge of people and listing of plants of a particular region are important tools that may help in understanding human-environment interactions. Consumption of wild edible plants meets the protein, carbohydrates, fats, vitamins, and mineral requirements of poor rural populace in the region. Wild edible plants are very important for the well-being of rural populations in the region, not only as sources of supplemental food, nutritionally balanced diets, medicines, fodder, and fuel, but also for their income-generating potential.

In conclusion, indigenous people constitute a large percentage of the global population. The areas where indigenous people live are very rich as far as the natural resources are concerned. In many places, they live close to the forest and depend on the forest to get food, fodder, and economy. Also, there is diversity in nutritional, anti-nutritional, and nutraceutical constituents among the wild edible plants consumed by the *Palliyars/ Kanikkars/Valliyans* and *Pulayans* in the Southeastern Slopes of Western Ghats. So new edible varieties with high nutritive values must be evolved by seeking recourse to modern bio-techniques. This will help the economically weaker sections of the population, including the tribal people living in various parts of India. To put it in a nutshell, this book will definitely shine light on the way forward to help initiate a number of development programs that have varied effects on this population.

Bibliography

Abasi, N. A., et al., (2013). Measuring efficiency of yam (Dioscorea spp.) production among resource-poor farmers in rural Nigeria. J. Agric. Food Sci. 1, 42–47.

Abhyankar, R. K., & Upadhyay, R., (2011). Ethnomedicinal studies of tubers of Hoshangabad, M. P. bulletin of environment. Pharmacol. Life Sci. 1, 57–59.

Acharya, S. N., & Thomas, J. E., (2007). Advances in medicinal plant research. Research Signpost, Kerala.

Achaya, K. T., (1998). A historical dictionary of Indian food. Oxford University Press, Delhi Cook BG, Pengelly BC, Brown SD, Donnelly JL, Eagles DA, Franco MA, Hanson J, Mullen BF, Partridge IJ, Peters M, Schultze-Kraft R (2005) Tropical Forages: an interactive selection tool. [CD-ROM], CSIRO, DPI&F(Qld). CIAT and ILRI, Brisbane.

Afolayan, A., & Jimoh, F. (2009). Nutritional quality of some wild leafy vegetables in South Africa. Int. J. Food Sci. Nutr. 60, 424–431.

Agbemafle, R. et al., (2012). Effects of boiling time on the concentrations of vitamin c and beta-carotene in five selected green vegetables consumed in Ghana. Adv. Appl. Sci. Res., 3, 2815–2820.

Aguiyi J. C. et al., (1997). Effects of *Mucuna pruriens* seed extract on smooth and skeletal muscle preparations. Fitoterapia. 68, 366–370.

Akhtar, M. S. et al., (1990). Antidiabetic evaluation of *Mucuna pruriens* Linn. seeds. JPMA. 40, 147–150.

Alexander, J., & Coursey, D. G., (1969). "The origins of yam cultivation," in The Domestication and Exploitation of Plants and Animals. Proceedings of a Meeting of the Research Seminar in Archaeology and Related Subjects Held at the Institute of Archaeology, London University, eds. P. J. Ucko and G. H. Dimbleby (Gerald: Duckworth & Co. Ltd.), 405–425.

Amadi, B. A. et al., (2013). Nutritional and Anti-Nutritional Evaluation of "Ji-otor" and "Ntubiri-ikpa" Traditional Foods of Ikwerre Ethnic National in Nigeria. J. Chem. Bio. Phy. Sci., 3, 1953–1962.

Amin, K. M. Y. et al., (1996). Sexual function improving effect of *Mucuna pruriens* in sexually normal male rats. Fitoterapia Milano. 67, 53–56.

Ana, H. et al., (2004). Pattern of use and knowledge of wild edible plants in distinct ecological environments: A case study of a Mapuche community from Northwestern Patagonia. Biodiversity Conserve. 13, 1153–1173.

Ananthakumar K. V. et al., (1994). Aphrodisiac activity of the seeds of *Mucuna pruriens*. Indian Drugs. 31, 321–327.

Anonymous. (1995). Ethnobiology in India: A status report, All India coordinated research project on ethnobiology (Ministry of Environment and Forests, Government of India, New Delhi).

AOAC., (2005). Official Methods of Analysis (18th edn.). Association of Official Analytical Chemists. Washington, DC.

APG III (Angiosperm Phylogeny Group); (2009). An update of the angiosperm phylogeny group classification for the orders and families of flowering plants: APG III. Bot. J. Linn. Soc. 161, 105–121.

Arinathan V, Mohan VR, De Britto AJ. Chemical composition of certain tribal pulses in south India. Int J Food Sci Nutr. 2003;54:209–217.

Arenas, P., & Scarpa, G., (2007). Edible wild plants of Chorote Indians, Gran Chacho, Argentina. Bot. J. Linn. Soc. 153, 73–85.

Arinathan, V. et al., (2007). Wild edibles used by Palliyars of the Western Ghats, Tamil Nadu, India. J. Trad. Knowl. 6, 163–168.

Arinathan, V. et al., (2009). Nutritional and antinutritional attributes of some under-utilized tubers. Trop. Subtrop. Agroecosyst. 10, 273–278.

Arnau, G. et al., (2010). “Yams” in Root and Tuber Crops, Handbook of Plant Breeding, ed. J. E. Bradshaw (Singapore: Springer Science Business Media, LLC), 127–148.

Arora, R, K., & Pandey, A., (1996). Wild edible plants of India, Diversity, Conservation and use. (Botanical Survey of India, Calcutta),

Arora, R. K., & Pandey Anjula, (1996). Ethnobotany in primitive tribe in Rajasthan. Wild edible plants of India: Diversity, Conservation and Use, (National Bureau of Plant Genetic Resources, New Delhi), 1.

Arora, R. K., (1991). Native food plants of the northeastern India. In: Contributions to Ethnobotany of India. (Ed) Jain, S. K. Scientific Publishers, Jodhpur, India. 137–152.

Asfaw, Z., & Tadesse, M., (2001). Prospects for sustainable use and development of wild food plants in Ethiopia. Econ. Bot. 55, 47–62.

Asfaw, Z., (1999). Ethnobotany of Nations, Nationalities and Peoples in Gambella, Benishangul-Gumuz and southern regions of Ethiopia. Research and Publication Office, Addis Ababa University.

Asfaw, Z., (2009). The future of wild food plants in Southern Ethiopia: Ecosystem conservation coupled with enhancement of the roles of key social groups. Acta Horticult. 806, 701–707.

Asiedu, R., et al., (1997). Yams, CGIAR Centers. Cambridge: Cambridge University Press.

Athena, D. P. et al., (2006). An ethnobotanical survey of wild edible plants of Paphos and Larnaca countryside of Cyprus. J. Ethnobiol. Ethnomed. 2, 34.

Auyeung, K. K. W., (2009). *Astragalus saponins* induce apoptosis via an ERK-independent NF-B signaling pathway in the human hepatocellular HepG2 cell line. Int. J. Mol. Med. 23, 189–196.

Ayensu, E. S., & Coursey, D. G., (1972). Guinea yams: the botany, ethnobotany, use and possible future of yams in West Africa. Econ. Bot. 26, 301–318.

Balemie, K., & Kibebew, F., (2006). Ethnobotanical study of wild edible plants in Derashe and Kucha Districts, South Ethiopia. J. Ethnobiol. Ethnomed. 2, 53.

Behera, K. K., (2006). Ethnomedicinal plants used by the tribals of Similipal Bioreserve Orissa, India: a pilot study. Ethnobot. Leaflet. 10, 149–173.

Belem, M. A. F., (1999). Application of biotechnology in the product development of nutraceuticals in Canada. Trends Food Sci. Tech. 10, 101–106

Bell, J., (1995). The hidden harvest. In seedling. The quarterly Newsletter of Genetic Resources Action International.

Beluhan, S., & Ranogajec, A., (2010). Chemical composition and non-volatile components of Croatian wild edible mushrooms. Food Chem. 124, 1076–1082.

Beta, T. et al., Sapirstein H. D. (2005). Phenolic content and antioxidant activity of pearled wheat and roller-milled fractions. Cereal Chem. 82, 390–393.

Bharucha, Z., & Pretty, J., (2010). The roles and values of wild foods in agricultural systems. Phil. Trans. R. Soc. B. 365, 2913–2926.

Bhatt, V. P., & Negi, G. C. S., (2006). Ethnomedicinal plant resources of Jaunsari tribe of Garhwal Himalaya, Uttaranchal. Indian J. Trad. Knowl. 5, 331–335.

Bhogaonkar, P. Y., & Kadam, V. N., (2006). Ethnopharmacology of Banjara tribe Umarkhed taluka, district Yavatamal, Maharashtra for reproductive disorder. Indian J. Tradit. Know. 5, 336–341.

Blagbrough, I. S. et al., (2010). Cassava: an appraisal of its phytochemistry and its biotechnological prospects—review. Phytochem. 711940–1951.

Brain, K. R., (1976). Accumulation of L-DOPA in cultures from *Mucuna pruriens*. Plant Science Lett. 7, 157–161.

Bravo, L. et al., (1998). Effect of various processing methods on the *in vitro* starch digestibility and resistant starch content of Indian pulses. J. Agric. Food. Chem. 46, 4667–4674.

Bravo, L. et al., (1999). Composition of underexploited Indian pulses. Comparison with common legumes. Food Chem. 64, 185–192.

Brink, M., (2006). *Macrotyloma uniflorum* (Lam.) Verdc. In: Brink M, Belay G (eds.) PROTA 1: cereals and pulses/céréales et legumes secs. PROTA, Wageningen.

Britta, M. et al., (2001). The contribution of Wild Vegetables to micronutrient intakes among women: An example from the Mekong Delta. Vietnam. Ecol. Food Nutr. 40, 159–184.

Britta, O. M. et al., (2003). Food, Feed or Medicine: The multiple functions of edible wild plants in Vietnam. Econ. Bot. 57, 103–117.

Britta, O. M., (2001). Wild vegetables and Micronutrient Nutrition studies on the Significance of Wild Vegetables in Women's Diets in Vietnam, (Comprehensive Summaries of Uppsala, Dissertations from the Faculty of Medicine).

Brower, V., (1998). Nutraceuticals: poised for a healthy slice of the healthcare market? Nat. Biotechnol. 16, 728–731.

Burkill, I. H., (1951). "Dioscoreaceae," in Flora of Malesiana, Vol. 4, ed. C. G. G. J. Van Steenis (Djakarta: Noordhoff), 293–335.

Burkill, I. H., (1960). The organography and the evolution of Dioscoreaceae, the family of the yams. J. Linn. Soc. (Botany). 56, 319–412.

Burns, R. B., (1971). Methods of estimation of tannin in the grain, sorghum. Agronomy J. 63, 511–512.

Cayen, M. N., & Dvornik, D., (1979). Effect of diosgenin on lipid metabolism in rats. J. Lipid Res. 20, 162–174.

Chan, Y. C. et al., (2010). Beneficial effects of yam on carbon tetrachloride-induced hepatic fibrosis in rats. J. Sci. Food Agric, 90, 161–167.

Chaudhri, R. D., (1996). Herbal drug industry: a practical approach to industrial pharmacognosy.

Chel-Guerrero, L. et al., (2002). Functional properties of flours and protein isolates from *Phaseolus lunatus* and *Canavalia ensiformis* seeds. J. Agric. Food Chem. 50, 584–591

Chen, H. L. et al., (2003). Effects of Taiwanese yam (*Dioscorea japonica* Thunb var. *pseudojaponica* Yamamoto) on upper gut function and lipid metabolism in Balb/c mice. Nutrit. 19, 646–651.

Chen, J. H. et al., (2008). Dioscorea improves the morphometric and mechanical properties of bone in ovariectomized rats. J. Sci. Food Agric. 88, 2700–2706.

Chen, Y. T., & Lin, K. W., (2007). Effects of heating temperature on the total phenolic compound, antioxidative ability and the stability of dioscorin of various yam cultivars. Food Chem. 101, 955–963.

Choi, E. M. et al., (2004). Immune cell stimulating activity of mucopolysaccharide isolated from yam (*Dioscorea batatas*). J. Ethnopharmacol. 91, 1–6.

Choudhary, K. et al., (2008). Ethnobotanical survey of Rajasthan: An update. Am. Eurasian J. Bot. 1, 38–45.

Chunekar, K. C., & Pandey, G. S., (1998). Bhavaprakash Nighantu (Indian Materia Medica) of Sri Bhavamisra (c. 1500–1600 AD). Chaukhamba Bharati Academy, Varanasi, 984.

Cook BG, Pengelly BC, Brown SD, Donnelly JL, Eagles DA, Franco MA, Hanson J, Mullen BF, Partridge IJ, Peters M, Schultze-Kraft R. Tropical Forages: an interactive selection tool. [CD-ROM], CSIRO, DPI&F(Qld) Brisbane: CIAT and ILRI; 2005.

Cooper, H. D., (1996). Promoting the identification, conservation and use of wild plants for food and agriculture in the Mediterranean: the FAO global plan of action. FAO, Rome.

Cornago, D. F. et al., (2011). Philippine Yam (Dioscorea spp.) tubers phenolic content and antioxidant capacity. Philippine J. Sci. 140, 145–152.

Coursey, D. G., (1967). Yams: An Account of the Nature, Origins, Cultivation and Utilization of the Useful Members of the Dioscoreaceae. London: Longmans, Greens and Co. Ltd.

D'Mello, J. P. F., (1995). Anti-nutritional substances in legume seeds. In: D'Mello J. P.F, Devendra C, editors. Tropical Legumes in Animal Nutrition, CAB International. Wallingford, UK, pp. 135–172.

Dansi, A. et al., (1999). Morphological diversity, cultivar groups and possible descent in the cultivated yams (*Dioscorea cayenensis*/*Dioscorea rotundata*) complex in Benin Republic. Genet. Resour. Crop Evol. 46, 371–388.

De Caluwé, E. et al., (2010a). *Adansonia digitata* L. – A review of traditional uses, phytochemistry and pharmacology. Afrika Focus. 23, 11–51, 53–83.

Debarata, D., (2004). Wild food plants of Madinapur, West Bengal used during Drought and Flood. Ethnobotany and Medicinal plants of India Subcontinent, Maheshwari, J. K. (ed.) Scientific Publishers Jodpur, India.

DeFeo, V., (1992). Traditional physiotherapy in peninsula Sorrentina, Campania, Southern Italy. J. Ethnopharmacol. 36, 113–125.

Deshmukh, S. R., & Rathod, V., (2013). Nutritional composition of wild edible *Ceropegia* tubers. Adv. Appl. Sci. Res., 4(1), 178–181.

Dhuley, J. N., & Naik. S. R., (1997). Protective effect of Rhinax, an herbal formulation against CCl_4 induced liver injury and survival in rats. J. Ethnopharm. 56, 159–164.

Di Patrizi, L. et al., (2006). Structural characterization of the N-glycans of gpMuc from *Mucuna pruriens* seeds.

Dickman, S. R., & Bray, R. H., (1940). Colorimetric determination of phosphate. Indust. Eng. Chem. Anal. Educ. 12, 665–668.

Dillard, C. J. and German, J. B. (2000). Phytochemicals: nutraceuticals and human health. J. Food Agric. Sci. 80, 1744–1756.

Dilworth, L. et al., (2012). Antioxidants, minerals and bioactive compounds in tropical staples. African J. Food Sci. Technol. 3, 90–98.

Donati, D. et al., (2005). Antidiabetic oligocyclitols in seeds of *Mucuna pruriens*. Phytotherapy Res. 19, 1057–1060.

Duke, J. A., (1981). Handbook of Legumes of World Economic Importance. New York, NY, USA: Plenum Press.

Dureja, H. et al., (2003). Developments in nutraceuticals. Indian J. Pharmacol. 35, 363–372.

Dutta, B., (2015). Food and medicinal values of certain species of *Dioscorea* with special reference to Assam. J. Pharmacog. Phytochem. 3, 15–18.

Edison, S. et al., (2006). Biodiversity of tropical tuber crops in India. Chennai: National Biodiversity Authority,

Egbung, G. E. et al., (2013). Chemical composition of root and stem bark extracts of *Nauclea latifolia*. Arch. Appl. Sci. Res., 5, 193–196.

Ertug, F., (2004). Wild edible plants of the Bodrum Area (Mugla, Turkey). Turkish J. Bot. 28, 161–174.

Etkin, N. L., & Ross, P. J., (1982). Food as medicine and medicine as food: an adaptive framework for the interpretation of plant utilization among the Hausa of northern Nigeria. Soc. Sci. Med. 16, 1559–1573.

FAO (Food and Agricultural Organization). (1993). The Sixth World Food Survey, FAO, United Nations, Rome. Reid, W. V., & Miller, K. R. (1989). Keeping the Options Alive: The Scientific Basis for Conserving Biodiversity. World Resource Institute, Washington D.C., USA.

FAO Publications; (1999). Use and potential of wild plants (Information Division, Food and Agriculture Organization of the United Nations, Rome, Italy).

FAO; Production Year Book, vol. 53, Food and Agriculture Organization, Rome, Italy, (1999).

FAO. (1989). Forestry and Nutrition, a reference manual. FAO Regional Office Bangkok.

FAO/WHO. (1991). Protein quality evaluation, (p 66). Rome, Italy: Food and Agricultural Organization of the United Nations.

FAOSTAT; (2013). http://faostat3.fao.org (accessed on 8 March 2019).

Feyssa, D., (2011). Wild edible fruits of importance for human nutrition in semi-arid parts of East Shewa Zone, Ethiopia: Associated indigenous knowledge and implications to food security. Pakistan J. Nutr. 10, 40–50.

Fleuret, A., (1993). Dietary and therapeutic uses of fruit in three Taita communities, in: Plants in Indigenous Medicine and Diet, by Etkin NL (Redgrave, Bedfort Hills, NY).

Food and Agriculture Organization (FAO). (1990). Roots, Tubers, Plantains and Bananas in Human Nutrition, vol. 24 of Food and Nutrition Series, Food and Agriculture Organization, Rome, Italy.

Ford-Lloyd, B. V. et al., (2011). Crop wild relatives undervalued, underutilized and under threat? BioSci. 61, 559–565.

Francesca, L., & Francesca, V., (2007). Wild food plants of popular use in Sicily. J. Ethnobiol. Ethnomed. 3, 15.

Gessler, M., & Hodel, U., (1997). In situ conservation of plant genetic resources in home gardens in Southern Vietnam (International plant genetic resources institute, Malaysia).

Ghorbani, A. et al., (2012). A comparison of the wild food plant use knowledge of ethnic minorities in Naban River Watershed National Nature Reserve, Yunnan, SW China. J. Ethnobiol. Ethnomed. 8, 17.

Giday, M. et al., (2009). Medicinal plants of the Meinit ethnic group of Ethiopia: An ethnobotanical study. J. Ethnopharmacol. 124, 513–521.

Gill, L. S., & Nyawuame, H. G. K., (1992). Leguminosae in ethnomedicinal practices of Nigeria. Ethnobotany. 6, 51–64.

Girach, R. D. et al., (1999). The medicinal flora of Similiphar forests, Orissa state, India. J. Ethnopharmacol. 65, 165–172.

Gireesha, J., & Raju, N. S. (2013). Ethnobotanical study of medicinal plants in BR Hills region of Western Ghats, Karnataka. Asian J. Plant Sci. Res., 3, 36–40.

Green, C., (1993). An overview of production and supply trends in the U. S. Specialty vegetable market. Acta Horticult. 318, 41–45.

Grivetti, L. E., & Britta, O. M., (2000). Value of traditional foods in meeting macro- and micronutrient needs: the wild plant connection. Nat. Res. Rev. 13, 31–46.

Guarrera, P. M., (2005). Traditional phytotherapy in Central Italy (Marche, Abruzzo, and Latium) Fitoterapia. 76, 1–25.

Guerranti, R. et al., (2001). Effects of *Mucuna pruriens* extract on activation of prothrombin by *Echis carinatus* venom. J. Ethnopharmacol. 75, 175–180.

Guerranti, R. et al., (2002). Proteins from Mucuna pruriens and enzymes from *Echis carinatus* venom: characterization and cross-reactions. J. Biol. Chem. 277, 17072–17078.

Guerranti, R. et al., (2004). Protection of *Mucuna pruriens* seeds against *Echis carinatus* venom is exerted through a multiform glycoprotein whose oligosaccharide chains are functional in this role. BBRC. 323, 484–490.

Guerranti, R. et al., (2008). Proteomic analysis of the pathophysiological process involved in the antisnake venom effect of *Mucuna pruriens* extract. Proteomics. 8, 402–412.

Gupta, M. et al., (1997). Antiepileptic and anticancer activity of some indigenous plants. Indian. J. Physiol. Allied Sci. 51, 53–56.

Gurumoorthi, P. et al., (2003). Nutritional potential of five accessions of a south Indian tribal pulse *Mucuna pruriens var. utilis*; II Investigation on total free phenolics, tannins, trypsin and chymotrypsin inhibitors, phytohaemagglutinins, and *in vitro* protein digestibility. Trop. Subtrop. Agroecosys. 1, 153–158.

Hahn, S. K., (1995). Yams, Dioscorea spp. (Dioscoreaceae), in Evolution of Crops Plants, eds J. Smartt and N. W. Simmonds (London: Longman Group Limited), 112–120.

Hamilton, A., (1995). The people and plant initiative. Px-xi. In G. J. Martin (ed.) Ethnobotany, Champan and Hall, London.

Haq, N., (1983). New food legume crop for the tropics. In: Better crops for food. (Eds.). Nugent, N. and Conor, M. O. Cuba Foundation Symposium 97. London, Pitman Books. pp. 144–160.

Harborne, J. B., (1994). Phytochemistry of the leguminosae. In: Bisby, F. A., Southon, I. W., editors. Phytochemical dictionary of the leguminosae. Boca Raton: Chapman and Hall.

Hasler, C. M., (2000). The changing face of functional food. J. Am. Coll. Nutr. 19, 499S–506S.

Hathcock, J., (2001). Dietary supplements: how they are used and regulated. J. Nutr. 131, 1114–1117.

Hazarika, T. K. et al., (2012). Studies on wild edible fruits of Mizoram, India used as ethnomedicine. Genet. Resour. Crop. Evol. 59, 1767–1776.

Heinrich, M. et al., (2005). 'Local food-nutraceuticals': an example of a multidisciplinary research project on local knowledge. J. Physiol. Pharmacol. 56, 5–22.

Heywood, V. H., (2011). Ethnopharmacology, food production, nutrition and biodiversity conservation: Towards a sustainable future for indigenous peoples. J. Ethnopharmacol. 137, 1–15.

Hishika, R. et al., (1981). Preliminary phytochemical and anti-inflammatory activity of seeds of *Mucuna pruriens*. Indian J. Pharmacol. 13, 97–98.

Horbovitz, M. et al., (1998). Fagopyritol B1, O-α-D-galactopyranosyl-(1→2)-D-chiro-inositol, a galactosylcyclitol in maturing buckwheat seeds associated with desiccation tolerance. Planta. 205, 1–11.

Hsu, C.K et al., (2011). Protective effects of the crude extracts from yam (*Dioscorea alata*) peel on tert-butylhydroperoxide-induced oxidative stress in mouse liver cells. Food Chem. 126, 429–434.

Hsu, F. L., (2002). Both dioscorin, the tuber storage protein of yam (*Dioscorea alata* cv. Tainong No. 1), and its peptic hydrolysates exhibited angiotensin converting enzyme inhibitory activities. J. Agric. Food Chem. 21, 6109–6113.

Humphries, E. C., (1956). Mineral composition and ash analysis In: Peach K. and M. V. Tracey (eds.) Modern Methods of Plant Analysis Vol.1, Springer-Verlag, Berlin, pp. 468–502.

Hussian, G., & Manyam, B. N., (1997). *Mucuna pruriens* proves more effective than L-Dopa in Parkinson's disease animal model. Phytother. Res. 11, 419–423.

Ilelaboye, N. O. A. et al., (2013). Effect of cooking methods on mineral and antinutrient composition of some green leafy vegetables. Arch. Appl. Sci. Res., 5, 254–260.

Issac, R. A., & Johnson, W. C., (1975). Collaborative study of wet and dry techniques for the elemental analysis of plant tissue by Atomic Absorption Spectrophotometer. J. Assoc. Off. Analyt. Chemist. 58, 376–338.

Jackson, M. L., (1967). Cyanide in plant tissue. In: Soil Chemical Analysis. Asia Publishing House New Delhi India. pp. 337.

Jadhav, V. D. et al., (2011). Documentation and ethnobotanical survey of wild edible plants from Kolhapur District. Rec. Res. Sci. Technol. 3, 58–63.

Jain, A., (2008). Some therapeutic uses of biodiversity among the tribal of Rajasthan. Indian J. Tradit. Know. 7, 256–262.

Jain, S. K., (Ed.). (1981). Glimpses of Indian Ethnobotany. Oxford & IBH Publishing Co., New Delhi, India.

Jana, S. K., & Chauhan, A. S., (1998). Wild edible plant of Sikkim Himalaya. J. Non-Timb. For. Prod. 5, 20–28.

Janardhanan, K., & Lakshmanan, K. K., (1985). Studies on the pulse, *Mucuna utilis*: Chemical composition and anti-nutritional factors. J. Food Sci. Technol. 22, 369–371.

Janardhanan, K. et al., (2003). Biodiversity of Indian underexploited/tribal pulses. In: Improvement Strategies for Leguminosae Biotechnology. (Eds.). Jaiwal, P. K. & Singh, R. P. Great Britain: Kluwer Academic Publishers. pp. 353–405.

Janardhanan, K. et al., (2003). Nutritional potential of five accessions of a South Indian tribal pulse, *Mucuna pruriens* var. *utilis*. Part I. The effect of processing methods on the contents of L-Dopa phytic acid, and oligosaccharides. Trop. Subtrop. Agroecosys. 1, 141–152.

Janardhanan, K., (1982). Studies on seed development and germination in *Mucuna utilis* Wall. ex. Wt. (Papilionaceae). PhD Thesis, Madras Univ., Madras, India.

Jasmine, T. S., (2007). Wild edible plants of Meghalaya, North- East India. Nat. Prod. Rad, 6, 410–426.

Javier, T. P. et al., (2006). Ethnobotanical review of wild edible plants in Spain botanical J. Linn. Soc. 152, 27–71.

Jeyaweera, D. M. A., (1981). Sri Lanka: National Science Council of Sri Lanka; Medicinal plants used in Ceylon Colombo.

Kadiri, M. et al., (2014). Ethnobotanical survey of plants commonly used for ceremonial activities among Yoruba tribe of South West Nigeria. Direct Res. J. Health Pharmacol. 2, 1–5.

Kala, K. B. et al., (2010). Nutritional and antinutritional potential of five accessions of a South Indian tribal pulse *Mucuna atropurpurea* DC. Tropical Subtrop. Agroeco. 12, 339–352.

Kalia, A. N., (2005). Textbook of Industrial Pharmacognocy, CBS publisher and distributor, New Delhi, 204–208.

Kalidass, C., & Mohan, V. R., (2011). Nutritional and antinutritional composition of itching bean (*Mucuna pruriens* (L.) DC var. *pruriens*): An underutilized tribal pulses in Western Ghats, Tamil Nadu. Tropical Subtrop. Agroeco. 14, 279–279.

Kamble, S. Y. et al., (2010). Studies on plants used in traditional medicines by Bhilla tribe of Maharashtra. Indian J. Tradit. Know. 9, 591–598.

Kanchan, L. V., (2011). Nutritional analysis of indigenous wild herbs used in eastern Chhattisgarh India. Emir. J. Food Agric. 23, 554–560.

Katewa, S. S., (2003). Contribution of some wild food plants from forestry to the diet of tribal of southern Rajasthan. Ind. Forester. 129, 1117–1131.

Khoshoo, T. N. (1991). Conservation of biodiversity in biosphere, In: Indian Geosphere-Biosphere Programme, Some aspects, National Academy of Sciences, Allahabad, India, 178–233.

Khyade, M. S. et al., (2009). Wild edible plants used by the tribes of Akole Tehsil of Ahmednagar District (MS), India. Ethnobot. Leaflets. 13, 1328–1336.

Kiremire, B. T., (2001). Indigenous food plants of Uganda. In: Proceedings of the 5th Colloquium of Natural Products Quebec, Canada, pp. 7–9.

Kokate, C. K. et al., (2002). Nutraceutical and Cosmaceutical. Pharmacognosy, 21st edition, Pune, India: Nirali Prakashan, 542–549.

Konsam, S., (2016). Erratum to: Assessment of wild leafy vegetables traditionally consumed by the ethnic communities of Manipur, Northeast India. J. Ethnobiol. Ethnomed. 12, 9.

Kulhalli, P., (1999). Heritage Healing. 29–30.

Kumar, D. S. et al., (2010). *In vitro* antioxidant activity of various extracts of whole plant of *Mucuna pruriens* (Linn). Int. J. Pharm. Tech. Res. 2, 2063–2070.

Kumar, S. et al., (2012). Study of wild edible plants among tribal groups of Similipal Biosphere Reserve forest, Odisha, India; with special reference to *Dioscorea* species. Int. J. Biol. Technol. 3, 11–19.

Kunkel, G., (1983). Plants for Human Consumption. Koeltz Scientific Books, Germany.

Kushi, L. H., (1999). Cereals, legumes, and chronic disease risk reduction: evidence from epidemiologic studies. Am. J. Clin. Nutr. 70, 451–458.

Lampariello, L. R. et al., (2012). The magic velvet bean of *Mucuna pruriens*. J. Tradit. Complement. Med. 2, 331–339.

Larner, J. et al., (1998). Phosphoinositol glycan derived mediators and insulin resistance Prospects for diagnosis and therapy. J. Basic Clin. Physiol. Pharmacol. 9, 127–137.

Lee, K. R. et al., (2004). Glycoalkaloids and metabolites inhibit the growth of human colon (HT29) and liver (HepG2) cancer cells. J. Agric. Food Chem. 52, 2832–2839.

Lev, L. S., & Shriver, A. L., (1998). "A trend analysis of yam production, area, yield, and trade (1961–1996)," in Ligname, Plante Seculaire et Culture Davenir. Actes du Seminaire International CIRAD-INRA-ORSTOM-CORAF (Montpellier: CIRAD), 8–10.

Li, B. W., & Cardozo, M. S., (1994). Determination of total dietary fiber in foods and products with little or no starch, non-enzymatic gravimetric method: collaborative study. J. Ass. Off. Analyt. Chem. Int. 77, 687–689.

Liu, Y. W. et al., (2007). Immunomodulatory activity of dioscorin, the storage protein of yam (*Dioscorea alata* cv. Tainong No. 1) tuber. Food Chem. Toxicol. 45, 2312–2318.

Lorenzetti, E., (1998). The phytochemistry, toxicology and food potential of velvet bean (*Mucuna adans* spp. Fabaceae) In: Buckles, D., Osiname, O., Galiba, M., Galiano G., editors. Cover crops of West Africa; contributing to sustainable agriculture. Ottawa, Canada & IITA, Ibadan, Nigeria: IDRC; p. 57.

Lulekal, E. et al., (2011). Wild edible plants in Ethiopia: A review on their potential to combat food insecurity. Africa Focus. 24, 71–121.

Ma, H. Y., (2002). Comparative study on anti-hypercholesterolemia activity of diosgenin and total saponin of *Dioscorea panthaica*. China J. Chinese Materia Medica. 27, 528–531.

Mahadkar, S. et al., (2013). Antioxidant activity of some promising wild edible fruits. Der Chemica Sinica, 4, 165–169.

Mahapatra, A. K., & Panda, P. C., (2012). Wild edible fruit diversity and its significance in the livelihood of indigenous tribals: Evidence from eastern India. Food Sec. 4, 219–234.

Maikhuri, R. K., (1991). Nutritional value of some lesser known wild food plants and their role in tribal nutrition: A case study in North-East India. J. Trop. Sci. 31, 397–405.

Maithili, V. et al., (2011). Antidiabetic activity of ethanolic extract of tubers of *Dioscorea alata* in alloxan-induced diabetic rats. Indian J. Pharmacol. 43, 455–459.

Majumdar, K. et al., (2006). Medicinal plants prescribed by different tribal and non-tribal medicine men of Tripura state. Ind. J. Trad. Know. 5, 559–562.

Malik, A., (2008). The potentials of Nutraceuticals. www.Pharmainfo.net, p. 6.

Mandal, P., (2005). Antimicrobial activity of saponins from *Acacia auriculiformis*. Fitoterapia. 76, 462–465.

Maneenoon, K. et al., (2008). Ethnobotany of *Dioscorea* L. (Dioscoreaceae), a major food plant of the Sakai tribe at Banthad Range, Peninsular, Thailand. Ethnobot. Res. Appl. 6, 385–394.

Mary Josephine, R., & Janardhanan, K., (1992). Studies on chemical composition and anti-nutritional factors in three germplasm seed materials of the tribal pulse, *Mucuna pruriens* (L.) DC. Food Chem. 43, 13–18.

Maxon, E. D., & Rooney, L. W., (1972). Two methods of tannin analysis for *Sorghum bicolor* (L.) Moench, grain. Crop Sci. 12, 253–254.

Mbiantcha, M. et al., (2011). Analgesic and anti-inflammatory properties of extracts from the bulbils of *Dioscorea bulbifera* L. var. *sativa* (Dioscoreaceae) in mice and rats. Evid-Based Compl. Alt. 912–935.

Meena, K. L., & Yadav, B. L., (2011). Some ethnomedicinal plants used by the Garasia tribe of district Sirohi, Rajasthan. Indian J. Tradit. Know. 10, 354–357.

Mehta, P. C., & Bhatt, K. C., (2007). Traditional soaps and detergent yielding plants of Uttaranchal. Indian J. Tradit. Know. 6, 279–284.

Menendez-Baceta, G. et al., (2012). Wild edible plants traditionally gathered in Gorbeialdea (Biscay, Basque Country). Genet. Resour. Crop Evol. 59, 1329–1347.

Mishra, S. et al., (2008). Wild edible tubers (*Dioscorea* spp.) and their contribution to the food security of tribes of Jaypore tract, Orissa, India. Plant Genet. Resour. 156, 63–67.

Misra, L., & Wagner, H., (2007). Extraction of bioactive principles from *Mucuna pruriens* seeds. Indian J. Biochem. Biophys. 44, 56–60.

Misra, R. C. et al., (2013). Genetic resources of wild tuberous food plants traditionally used in Similipal Biosphere Reserve, Odisha. India. Genet. Resour. Crop Evol. 60, (2033).

Mohan, V. R., & Janardhanan, K., (1995). Chemical analysis and nutritional assessment of lesser-known pulses of the genus. *Mucuna*. Food Chem. 52, 275–280.

Moreno-Black et al., (1996). Cultivating continuity and creating change: women's home garden practices in northeastern Thailand. Agric. Human Values. 1, 33–11.

Muller, H. G., & Tobin, G., (1980). Nutrition and Food Processing, London: Croom Helm Ltd .

Muralidharan, P. K. et al., (1997). Biodiversity in Tropical Moist Forest: Study of Sustainable Use of Non-Wood Forest Products in the Western Ghats. KFRI Research Report No.133. KFRI, Thrissur, Kerala, India.

Murugkar, D. A., & Susbulakhmi, G., (2005). Nutritive values of wild edible and species consumed by the Khasi tribe of India. Ecol. Food Nutr. 44, 207–223.

Nag, A., (1999). A Study of the Contribution of Some Wild Food Plants to the Diet of Tribals of South East Rajasthan. PhD thesis, Mohanlal Sukhadia University, Udaipur.

Narzary, H. et al., (2013). Wild Edible Vegetables Consumed by Bodo Tribe of Kokrajhar District (Assam), North- East India. Arch. Appl. Sci. Res., 5, 182–190.

Nashriyah, M. et al., (2011). Ethnobotany and distribution of wild edible tubers in Pulau Redang and nearby islands of Tereengganu, Malaysia. Int. J. Biol. Vert. Agric. Food Eng. 5, 110–113.

Nassar, N. M. A. et al., (2008). Wild Manihot species: botanical aspects, geographic distribution and economic value. Gen.Mol. Res. 7, 16–28.

Nayaboga, E. et al., (2014). Agrobacterium-mediated genetic transformation of yam (*Dioscorea rotundata*): an important tool for functional study of genes and crop improvement. Front. Plant Sci. 5, 463.

Nayak, S. et al., (2004). Ethno-medico botanical survey of Kalahandi District of Odisha. Indian J. Tradit. Know. 3, 72–79.

Neelam, D. A., (2007).Identification and quantification of nutraceuticals from Bengal gram and horse gram seed coat. Dissertation for Bachelor of Technology. Department of Biotechnology, Sathyabama University Chennai (India), India.

Ogbonna, O. J. et al., (2013). Comparative studies of the phytochemical and proximate analysis; mineral and vitamin compositions of the root and leaf extracts of *Tetracarpidium conophorum*. Arch. Appl. Sci. Res., 5, 55–59.

Ogundare, A. O., Olorunfemi, O. B., (2007). Antimicrobial efficacy of the leale of *Dioclea reflexa, Mucana pruriens, Ficus asperifolia, and Tragia spathulata*. Res. J. Microbiol. 2, 392–396.

Oko, M. O. et al., (2012). The Effects of Ethanol Extract of *Allium sativum* Leaves on Aspartate Aminotransferase, Alanine Aminotransferase, and Alkaline Phosphatase in Albino Rats. J. Chem. Bio. Phy. Sci., 3, 256–263.

Olayemi, J. O., & Ajaiyeoba, E. O., (2007). Anti-inflammatory studies of yam (*Dioscorea esculenta*) extract on Wistar rats. African J. Biotech. 6, 1913–1915.

Omar, N. F. et al., (2012). Phenolics, flavonoids, antioxidant activity and cyanogenic glycosides of organic and mineral-based fertilized cassava tubers. Molecules. 17–2378–2387.

Ortmeyer, H. K. et al., (1995). Effect of D-chiro-inositol added to a meal on plasma glucose and insulin in hyperinsulinemic rhesus monkeys. Obesity Res. 3, 605S–608S.

Padmaja, G. et al., (2001). Digestibility of Starch and Protein. Thiruvananthapuram: Central Tuber Crops Research Institute,

Pandey, A. et al., (2008). Towards collection of wild relatives of crop plants in India. Genet. Resour. Crop Evol. 55, 187–202.

Pandey, M. et al., (2010). Nutraceuticals: new era of medicine and health. Asian J. Pharm. Clin. Res. 3, 11–15.

Paśko, P. et al., (2009). Anthocyanins., total polyphenols and antioxidant activity in amaranth and quinoa seeds and sprouts during their growth. Food Chem. 115, 994–998

Pastor, A., & Gastavo, F. S., (2007). Edible wild plants of the Cherote India, Gran chalo, Argentina. Bot. J. Linn. Soc. 153, 73–85.

Patil, M. V., & Patil, D. A., (2005). Ethnomedicinal practices of Nasik district, Maharashtra. Indian J. Tradit. Know. 4, 287–290.

Patwardhan, B. et al., (2005). Ayurveda and traditional Chinese medicine: a comparative overview. eCAM, 2, 465–473.

Perumal, S., & Sellamuthu, M. (2007). The antioxidant activity and free radical scavenging capacity of dietary phenolic extracts from horse gram (*Macrotyloma uniflorum* (Lam.) Verdc.) seeds. Food Chem. 105, 950–958

Pfoze, N. L. et al., (2012). Assessment of local dependency on selected wild edible plants and fruits from senapati district, Manipur, Northeast India. Ethnobot. Res. Appl. 10, 357–367.

Pieroni, A. et al., (2002). Ethnopharmacology of Liakra: traditional weedy vegetables of the Arbëreshë of the Vulture area in southern Italy. J. Ethnopharmacol. 81, 165–185.

Pilgrim, S. et al., (2008). Ecological knowledge is lost in wealthier communities and countries. Environ. Sci. Tech. 42, 1004–1009.

Polhill, R. M., & Raven, P. H., (1981). Advances in Legume Systematics. Kew: Royal Botanic Gardens.

Pramila, S. S. et al., (1991). Nutrient composition of some uncommon foods consumed by Kumaon and Garhwal hill subjects. J. Food Sci. Technol. 28, 237–238.

Prasad, K. et al., (2010). Compositional characterization of traditional medicinal plants: Chemo-metric approach. Arch. Appl. Sci. Res., 2, 1–10.

Prawat, H. et al., (1995). Cyanogenic and non-cyanogenic glycosides from *Manihot esculenta*. Phytochem. 401167–1173.

Pugalenthi, M. et al., (2005). Alternative food/feed perspectives of an under-utilized legume *Mucuna pruriens* Utilis: A Review. Linn J. Plant Foods Human Nutr. 60, 201–218.

Purseglove, J. W., (1974). *Dolichos uniflorus*. In: Tropical Crops: Dicotyledons, Longman, London, 263–264.

Radhakrishnan, K. et al., (1996). Ethnobotany of the wild edible plants of Kerala, India. In: Ethnobotany in Human Welfare (Ed.) Jain SK. Deep Publications, New Delhi, India, pp. 48–51.

Rajendran, V. et al., (1996). Reappraisal of dopamineric aspects *Mucuna pruriens* and comparative profile with L-DOPA on cardiovascular and central nervous system in animals. Indian drugs. 33, 465–472.

Rajurkar, N. S., & Gaikwad, K., (2012). Evaluation of phytochemicals, antioxidant activity and elemental content of *Adiantum capillus veneris* leaves. J. Chem. Pharm. Res., 4, 365–374.

Rajyalakshmi, P., & Geervani, P., (1994). Nutritive value of the foods cultivated and consumed by the tribals of South India. Plant Food. Hum. Nutr. 46, 53–61.

Rakesh, K. M. et al., (2004). Bioprospecting of wild edible for rural development in the central Himalaya Mountains of India. Mountain Res. Dev. 24, 110–113.

Rapheal, O., & Adebayo, K. S., (2011). Assessment of trace heavy metal contaminations of some selected vegetables irrigated with water from River Benue within Makurdi Metropolis, Benue State Nigeria. Adv. Appl. Sci. Res., 2, 590–601.

Rashid, A., (2008). Less known wild edible plants used by the Gujjar tribe of District Rajouri, Jammu & Kashmir state India. Int. J. Bot. 4, 219–224.

Raskin, I. et al., (2002). Plants and human health in the twenty-first century. Trends. Biotechnol. 20, 522–531.

Ravindran, V., & Ravindran, G., (1988). Nutritional and antinutritional characteristics of *Mucuna* (*Mucuna utilis*) bean seeds. J. Sci. Food Agric. 46, 71–79.

Ravishankar, K., & Vishnu Priya, P. S., (2012). *In vitro* antioxidant activity of ethanolic seed extracts of *Macrotyloma uniflorum* and *Cucumis melo* for therapeutic potential. Int. J. Pharmacy Chem. 2, 442–445.

Rawat, S. W. et al., (1994). Biochemical investigation of some common wild fruits in Garhwal Himalaya. Prog. Hortic. 26, 35–40.

Reilly, K. et al., (2004). Oxidative stress responses during cassava post-harvest physiological deterioration. Plant Mol. Biol. 56–625–641.

Reyes-Garcia, V. et al., (2005). Knowledge and consumption of wild plants: A comparative study in two Tsimane villages in the Bolivian Amazon. Ethnobot. Res. App. 3, 201–207.

Rishi, R. K., (2006). Nutraceuticals: borderline between food and drug? Pharma. Rev. 51–53.

Rout, S. D., & Panda, S. K., (2010). Ethnomedicinal plant resources of Mayurbhanj district, Orissa. Indian J. Tradit. Know. 9, 68–72.

Roy, B. et al., (1998). Plants for Human Consumption in India. Botanical Survey of India, Calcutta.

Sadasivam, S., & Manickam, A., (1996). Biochemical methods, New Age International (P) Limited Publishers, New Delhi, India.

Sahu, S. C. et al., (2010). Potential medicinal plants used by the tribal of Deogarh district, Orissa, India. Ethanol. Med. 4, 53–61.

Sahu, T. R., (1996). Life support promising food plants among aboriginal of Baster (M.P), India. In: Ethnobiology in Human Welfare (ed.) Jain, S. K. New Delhi, India: Deep Publications. pp. 26–30.

Saikia, P., & Deka, D. C., (2013). Mineral content of some wild green leafy vegetables of North-East India. J. Chem. Pharm. Res., 5, 117–121.

Samanta, A. K., & Biswas, K. K., (2009). Climbing plants with special reference to their medicinal importance from Midnapore Town and its adjoining areas. J. Econom. Taxonom. Bot. 33, 180–188.

Sami Labs—Pioneer in Nutraceuticals. The Hindu Newspaper; Aug 5, 2002.

Sarmah, B. P. et al., (2013). Wetland medicinal plants in floodplains of Subansiri and Ranga river of Lakhimpur district, Assam, India. Asian J. Plant Sci. Res., 3, 54–60.

Sasi, R., & Rajendran, A., (2012). Diversity of wild fruits in Nilgiri hills of the southern Western Ghats ethnobotanical aspects. Int. J. App. Biol. Pharm.Tech. 3, 83–87.

Sasi, R. et al., (2011). Wild edible plant Diversity of Kotagiri Hills – a Part of Nilgiri Biosphere Reserve, Southern India. J. Res. Biol. 2, 80–87.

Sathiyanarayanan, L., & Arulmozhi, S., (2007). *Mucuna pruriens* A comprehensive review. Pharmacognosy Rev. 1, 157–162.

Seal, T., (2012). Evaluation of nutritional potential of wild edible plants, traditionally used by the tribal people of Meghalaya state in India. Amer. J. Plant Nutr. Fertil. Tech. 2, 19–26.

Setalaphruk, C., & Lisa, L. P., (2007). Children's traditional ecological knowledge of wild food resources: a case study in a rural village in Northeast Thailand. J. Ethnobiol. Ethnomed. 3, 1–11.

Shajeela, P. S. et al., (2011). Nutritional and anti-nutritional evaluation of Wild Yam (*Dioscorea* spp.). Trop. Subtrop. Agroecosyst. 14, 723–730.

Shang, H. F. et al., (2007). Immunostimulatory activities of yam tuber mucilages. Bot. Stud. 48, 63–70.

Shanthakumari, S. et al., (2008). Nutritional evaluation and elimination of toxic principles in wild yam (*Dioscorea* spp.). Trop. Subtrop. Agroecosyst. 8, 313–319.

Sharma, L. N., & Bastakoti, R., (2009). Ethnobotany of Dioscorea L. with emphasis on food value in Chepang communities in Dhading District, central Nepal. Botanica Orientalis. J. Plant Sci. 6, 12–17.

Shaw, B. P., & Bera, C. H., (1993). A preliminary clinical study to evaluate the effect of Vigorex-SF in sexual debility patients. Indian J. Inter Med. 3, 165–169.

Sheikh, N., (2013). Phytochemical screening to validate the ethnobotanical importance of root tubers of *Dioscorea* species of Meghalaya, North East India. J. Med. Plant Stud. 1, 62–69.

Siddhuraju, P., & Becker, K., (2001). Rapid reversed-phase high performance liquid chromatographic method for the quantification of L-Dopa (L-3, 4-dihydroxyphenylalanine), non-methylated and methylated tetrahydroisoquinoline compounds from *Mucuna* beans. Food Chem. 72, 389–394.

Siddhuraju, P. et al., (1996a). Chemical composition and protein quality of the little-known legume, velvet bean (*Mucuna pruriens* [L.] DC.). J. Agric. Food Chem. 44, 2636–2641.

Singh, H. B., & Arora, R. K., (1978). Wild Edible Plants of India, Indian Council of Agricultural Research (ICAR), New Delhi.

Singh, H. B. et al., (1997). Ethnomedico botanical studies in Tripura, India. Ethnobot. 9, 56–58.

Singh, N. et al., (2009). Some ethnobotanical plants of Ranikhet region, Uttaranchal. J. Econ. Taxon. Bot. 33, 198–204.

Sinha, R., & Lakra, V., (2005). Wild tribal food plants of Orissa. Indian J. Tradit. Know. 4, 246–252.

Slikkerveer, L., (1994). Indigenous agricultural knowledge systems in developing countries: a bibliography project report on Indigenous Knowledge Systems Research and Development Studies No. 1 Special issue: INDAKS in collaboration with the European Commission DG XII Leiden, the Netherlands,

Sodani, S. N. et al., (2004). Phenotypic stability for seed yield in rainfed Horse gram (*Macrotyloma uniflorum* [Lam.] Verdc). Paper presented in National Symposium on Arid Legumes for Sustainable Agriculture and Trade, 5–7 Nov. 2004. Central Arid Zone Research Institute, Jodhpur.

Sofowora, A., (1982). Medicinal Plants in Traditional Medicine in West Africa, 1st Edn. London: John Wiley and Sons.

Son, I. S. et al., (2007). Antioxidative and hypolipidemic effects of diosgenin, a steroidal saponin of yam (*Dioscorea* spp.), on high-cholesterol-fed rats. Biosci. Biotechnol. Biochem. 71, 3063–3071.

Sonibare, M. A., (2012). In vitro antimicrobial and antioxidant analysis of *Dioscorea dumetorum* (Kunth) Pax and *Dioscorea hirtiflora* (Linn.) and their bioactive metabolites from Nigeria. J. App. Biosci. 51, 3583–3590.

Spencer, J. P. E. et al., (1996). Evaluation of the pro-oxidant and antioxidant actions of L-Dopa and dopamine *in vitro*: implications for Parkinson's disease. Free Rad. Res. 24, 95–105.

Spencer, J. P. E. et al., (1995). Superoxide-dependent GSH depletion by L-Dopa and dopamine relevance to Parkinson's disease. Neuroreport. 6, 1480–1484.

Spina, M. B., & Cohen, G., (1988). Exposure of school synaptosomes to L-Dopa increases levels of oxidized glutathione. J. Pharmacol. Exp. Ther. 247, 502–507.

Su, P. F. et al., (2011). *Dioscorea* phytocompounds enhance murine splenocyte proliferation *ex vivo* and improve regeneration of bone marrow cells *in vivo*. Evidence-Based Complement. Altern. Med. Article ID 731308, 11 pp.

Sundriyal, M., & Sundriyal, R. C., (2001). Wild edible plants of the Sikkim Himalaya: Nutritive values of selected species. Econ. Bot. 55, 377–390.

Sundriyal, M. et al., (1998). Wild edibles and other useful plants from the Sikkim Himalaya, India. Oecologia Montana. 7, 43–54.

Swarnkar, S., & Katewa, S. S., (2008). Ethnobotanical Observation on Tuberous Plants from tribal area of Rajasthan (India). Ethnobot. Leaflet. 12, 647–666.

Tamiru, M. et al., (2008). Diversity, distribution and management of Yam landraces (*Dioscorea* spp.) in Southern Ethiopia. Genet. Resour. Crop Evol. 55, 115–131.

Tardio, J. et al., (2006). Ethnobotanical review of wild edible plants in Spain. Bot. J. Linn. Soc. 152, 27–71.

Teklehaymanot, T., & Giday, M., (2010). Ethnobotanical study of wild edible plants of Kara and Kewego semipastoralist people in Lower Omo River valley, Debub Omo Zone, SNNPR, Ethiopia. J. Ethnobiol. Ethnomed. 6, 23.

Temel, R. E. et al., (2009). Diosgenin stimulation of fecal cholesterol excretion in mice is not NPC1L1 dependent. J. Lipid Res. 50, 915–923.

Termote, C. et al., (2010). Eating from the wild. Turumbu indigenous knowledge on non-cultivated edible plants, district Tshopo, DR. Congo. Ecol. Food Nutr. 49, 173–207.

Termote, C. et al., (2011). Eating from the wild. Turumbu, Mbole and Bali traditional knowledge on non-cultivated edible plants, Tshopo district, DR. Congo. Ecol. Food Nutr. 49, 173–207.

Termote, C., (2009). Use and socioeconomic importance of wild edible plants in tropical rainforest around Kisangani district, Tshopo, DR Congo. In: Systematics and conservation of African plants (X. Van der Burgt, J. Van der Maesen and J. M. Onana eds.). pp. 415–425. Royal Botanic Gardens, Kew.

Teron, R., (2011). Studies on Ethnobotany of Karbi-Anglong District, Assam: Transcultural Dynamism in Traditional Knowledge. PhD thesis, Guwahati University, Assam.

Thewles, A. et al., (1993). Effect of diosgenin on biliary cholesterol transport in the rat. Biochem. J. 291, 793–798.

Tiwari, L., & Pande, P. C., (2006). Indigenous veterinary practices of Darma Valley of Pithoragrah district, Uttaranchal. Indian J. Tradit. Know. 5, 201–206.

Tripathi, A. S. et al., (2010). Immunomodulatory activity of the methanol extract of *Amorphophallus campanulatus* (Araceae) Tuber. Trop. J. Pharma. Res. 9, 451–454.

Tripathi, Y. B., & Updhyay, A. K., (2001). Antioxidant property of *Mucuna pruriens*. Linn. Curr. Sci. 80, 1377–1378.

Uchida, K. et al., (1984). Effects of diosgenin and B-sisterol on bile acids. The J. Lipid Res. 25, 236–245.

Ujowundu, C. O. et al., (2010). Evaluation of the chemical composition of *Mucuna utilis* leaves used in herbal medicine in Southeastern Nigeria. Afr. J. Pharm. Pharmacol. 4, 811–81.

Uprety, Y. et al., (2012). Diversity of use and local knowledge of wild edible plant resources in Nepal. J. Ethnobiol. Ethnomed. 8, 16.

Vadivel, V., & Janardhanan, K., (2000). Preliminary agrobotanical traits and chemical evaluation of *Mucuna pruriens* (Itching bean): a less-known food and medicinal legume. J. Med. Aromatic Plant Sci. 22, 191–199.

Vadivel, V., & Pugalenthi, M., (2008). Removal of antinutritional/toxic substances and improvement in the protein digestibility of velvet bean seeds during various processing methods. J. Food Sci. Technol. 45, 242–246.

Vasudeva Rao, M. K., & Shanpru, R., (1991). Some plants in the life of the Garos of Meghalaya. In: Contributions to Ethnobotany of India. (Ed.) Jain S.K, Jodhpur, India: Scientific Publishers. pp. 183–190.

Victoria, R. G. et al., (2006). Cultural practical and economical value of wild plants: A quantitative study in Bolivian amazon. Econ. Bot. 60, 62–71.

Wanasundera, J. P., & Ravindran, G., (1994). Nutritional assessment of Yam (*Dioscorea alata*) tubers. Plant Food Hum. Nutr. 46, 33–39.

Wang, T. S. et al., (2011). Anticlastogenic effect of aqueous extract from water yam (*Dioscorea alata* L.). J. Medicinal Plant Res. 5, 6192–6202.

Williams, D. E., (1993). *Lyanthes moziniana* (Solanaceae): An underutilized Mexican food plant with 'new' crop potential. Eco. Bot. 47, 387–400.

Wilson, K. B., (1990). Ecological dynamics and human welfare: a case study of population, health and nutrition in southern Zimbabwe, PhD thesis, University College, London.

Wu, W. H. et al., (2005). Estrogenic effect of yam ingestion in healthy postmenopausal women. J. Amer. Coll. Nut. 24, 235–243.

Yadava, N. D., & Vyas, N. L., (1994). Arid legumes. Agrobios, India.

Zemede, A., (1997). Indigenous African food crops and useful plants: survey of indigenous crops, their preparations and home gardens. NAIROBI: The United Nation University institute for Natural Resources in Africa.

Zhi-Dong, X., (2009). The differentiation and proliferation inhibitory effects of sporamin from sweet potato in 3T3-L1 preadipocytes. Agric. Sci. China. 8, 671–677.

Index